A Guide to

Managing Indoor Air Quality

in Health Care Organizations

Edited by Wayne Hansen, PE, REA, CEM

Joint Commission Mission

The mission of the Joint Commission on Accreditation of Healthcare Organizations is to improve the quality of care provided to the public through the provision of health care accreditation and related services that support performance improvement in health care organizations.

Joint Commission educational programs and publications support, but are separate from, the accreditation activities of the Joint Commission. Attendees at Joint Commission educational programs and purchasers of Joint Commission publications receive no special consideration or treatment in, or confidential information about, the accreditation process.

About the Editor:

Wayne Hansen, PE, REA, CEM, is Director of Engineering, Mintie Corporation, headquartered in Los Angeles, California. He has been a registered engineer for more than 25 years, and has specialized in indoor air quality assessments relative to energy conservation for more than 15 years. He currently serves on the Indoor Air Quality Advisory Board for California OSHA.

About the Contributors:

Fred Brunsmann is Chief Engineer, Kaiser Foundation Hospital, Panorama City, California. He is a plant engineer with more than 38 years of experience in facilities maintenance and operation. He has been with Kaiser for more than 25 years.

H. E. Barney Burroughs, CIAQP, is Director and Principal, Environmental Design International, Ltd, Alpharetta, Georgia. He is a consultant with more than 35 years of experience in the field of mechanical system design, filtration and air cleaning, and indoor air quality analysis. He is a presidential member and fellow of ASHRAE. He is also the chairman of the IAQ certification Board of the Association of Energy Engineers.

Stephen M. Burch, PE, is Principal Engineering Consultant, Honeywell Inc, Healthcare Business Unit, Nashville. He is a registered engineer with more than 16 years of design and construction experience in the health care field. He is an active member of the American Hospital Association and the American Society for Healthcare Engineering.

Brian Krafthefer, PE, DEE, is Staff Research Engineer, Honeywell Technology Center, Minneapolis, Minnesota. He is a registered engineer with an educational background in mechanical engineering and physics. He has more than 18 years of experience in the research and application of HVAC systems and IAQ.

Charlotte G. Lee, RN, CIC, is Director of Infection Control and Epidemiology, Kaiser Permanente, Woodland Hills, California. She has more than 25 years of experience in the nursing and infection control fields, and 14 years of experience in occupational health and safety.

Kenneth E. Lewis, MPA, is Safety Officer, Scottsdale Memorial Hospitals, Scottsdale, Arizona. He directs the safety and industrial hygiene programs at the hospital. In addition, he has nearly 20 years of experience in hazardous materials management and oversees the corporatewide Hazardous Materials and Waste Management Program.

Bud Maxfield is Director of Facilities, Torrance Memorial Medical Center, Torrance, California. He has a degree in engineering, and more than 35 years of experience in facilities engineering and construction management. He has been with TMMC for 17 years, nine years as Construction Manager, and eight as Director of Facilities.

John F. McCarthy, ScD, CIH, is President, Environmental Health & Engineering, Inc, Newton, Massachusetts. He is a recognized expert in the field of indoor air quality and a leader in identifying emerging issues. He serves on the Indoor Air Quality Engineering Research Subcommittee for the EPA's Science Advisory Board, among other committees, and is widely published in the areas of exposure assessment, control measures, and program development.

Philip R. Morey, PhD, CIH, is Director of Microbiology and IAQ and Vice President, AQS Sciences, Gettysburg, Pennsylvania. He has been actively involved in the industrial hygiene community, and in the bioaerosol and indoor air quality fields for more than 30 years. He is a member of the International Academy of Indoor Air Sciences and Co-chair of the ISIAQ task force on control of moisture problems affecting biological IAQ.

Irene Orme, RN, CIC, is an independent consultant based in West Hills, California. She has been active in nursing for more than 35 years. She has more than 20 years of experience in dealing with regulatory issues in the field of infection control and epidemiology. She has twice served as a president of the Greater Los Angeles Chapter of the Association of Practitioners in Infection Control and Epidemiology (APIC), and has chaired the California APIC Coordinating Council.

Anand K. Seth, PE, CEM, is the Corporate Director of Facilities Engineering for Partners Health Care Systems Inc, of Boston, Massachusetts. Partners is an integrated health care system including the Massachusetts General Hospital, Brigham and Women's Hospital, and North Shore Medical Center.

Andrew J. Streifel, MPH, REHS, is Environmentalist, Department of Environmental Health & Safety, University of Minnesota, Minneapolis. Mr Streifel has more than 20 years of experience as the Hospital Environment Specialist for the Department of Environmental Health and Safety at the University and as adjunct faculty in the School of Public Health. Mr Streifel has been widely published in professional journals and in newspaper articles regarding his experience in more than 150 health care facilities relating to indoor air quality issues.

Contents

A Guide to Managing Indoor Air Quality in Health Care Organizations

Part I: Identifying the Issues

Foreword

Anand K. Seth PE, CEM, *is the Corporate Director of Facilities Engineering for Partners HealthCare Systems Inc, of Boston, Massachusetts. Partners is an integrated health care system including the Massachusetts General Hospital, Brigham and Women's Hospital, and North Shore Medical Center.*

As the complexity of building construction, particularly HVAC systems and their controls, has increased, many indoor air quality (IAQ) problems have arisen, including sick-building syndrome (SBS) and building-related illness (BRI). Addressing these problems requires time and energy, not to mention money: Recent Environmental Protection Agency (EPA) estimates show that in the United States alone, the effects of poor IAQ can cost $40 billion per year in lost worker productivity and $1.2 billion in annual medical costs. These problems are not limited to the United States; everywhere in the world where central HVAC systems are used for buildings, IAQ problems have been noted. The issue of poor IAQ has created a worldwide demand for understanding and remediation solutions.

In the 1970s, to meet demands for cost-containment, large facilities reduced their ventilation rates and minimized their operating and maintenance practices. This affected IAQ, as airborne fungi and bacteria proliferated, resulting in such highly publicized cases as the Legionnaire's disease incident. Improper engineering, construction, and systems installation, performed with little or no commissioning, added to the problem. We are just beginning to understand the impact of contaminants (such as mold and the off-gassing from various synthetic materials now commonly used in buildings) on our indoor environment.

However, as a cost-effective remedial solution, just increasing ventilation (outside air) is usually not sufficient to improve IAQ. In fact, high ventilation rates without proper humidity controls can be *more* problematic. For instance, in the relatively dry northern climates of the United States, a high ventilation rate without humidification causes extremely low humidity in the conditioned space. The reverse is true in the warmer, humid climates of the South, Southeast, and Hawaii, where high ventilation rates have caused excessively high relative humidity and a corresponding increase in fungi-related IAQ problems.

Health care facilities, by their nature, pose unique IAQ challenges. A diverse population is affected, including patients, visitors, and health care workers. Furthermore, HVAC system maintenance is a constant issue. Because health care facilities are complex and subjected to severe and changing operational demands, decontamination and infection control are continual problems—but continuous operation and 24-hour demands on HVAC systems make it difficult to find adequate opportunity for proper maintenance.

This manual is an excellent and timely guide to understanding and remediating IAQ issues in health care facilities. The contributors present IAQ theories and practical applications in an easily understood style that clarifies the economic and human costs of IAQ and addresses the need to avoid IAQ problems at the outset or confront them head-on when they arise.

Preface

Indoor air quality is a topic that is highly publicized in today's regulatory environment, and implications for health care organizations must be carefully evaluated. IAQ issues would generally fall under Joint Commission standards relating to utility systems and infection control, although indoor air quality is not specifically addressed in any Joint Commission standard.

This publication is meant to serve as a guide for an indoor air quality evaluation process and to provide examples and possible solutions to IAQ issues.

Introduction

The Need for an Integrated Indoor Air Quality Program

Wayne Hansen, PE, REA, CEM, *is Director of Engineering, Mintie Corporation, headquartered in Los Angeles, California. He has been a registered engineer for more than 25 years, and has specialized in indoor air quality assessments relative to energy conservation for more than 15 years. He currently serves on the Indoor Air Quality Advisory Board for California OSHA.*

Indoor air quality (IAQ) is a convenient umbrella term that encompasses sick building syndrome (SBS), building-related illness (BRI), and indoor air pollution (IAP). The purpose of this book is to provide personnel in the departments of facility management, employee safety, and infection control with an understanding of IAQ issues and a painless method of integrating a comprehensive IAQ program into their organizations. Although IAQ as a topic is relatively new, attention from all quarters and all industries is increasing. Because of health care facilities' special position in the community, the need for a sound IAQ program as part of a facility's vital utilities management plan is profound.

Often, when IAQ in health care facilities is discussed, the focus is on patient safety. In contrast, the focus of this book is on the health care professionals and facility employees because their exposure to sources of potential airborne contamination is much greater, due to their constant exposure to anything harmful in a health care facility. Of course, unhealthful air originating in the workplace affects all building occupants; but real or perceived poor IAQ in a health care organization can have a serious effect on employee health and morale, not to mention the financial health of the organization.

In 1994, the Occupational Safety and Health Administration (OSHA) introduced a performance-based set of proposed regulations governing indoor air quality.[1] This proposal was the result of several years of study of this issue by the National Institute of Occupational Safety and Health (NIOSH), a research sibling organization to OSHA, and by the Environmental Protection Agency (EPA). The complete text of these regulations was published in the April 5, 1994, issue of the Federal Register to solicit public comment and testimony. OSHA underestimated the extent of the interest that the public would have in this proposal. By March 13, 1995, they had received more than 110,000 comments and testimony that extended the original comment period by many months. This

Element	Specific Actions
Program Establishment	1. Design the program. 2. Designate a person in charge. 3. Gather all facility-related plans and documents together.
Program Implementation	4. Keep a written log of IAQ complaints and disposition. 5. Follow good safety practices with regard to noxious materials and employee safety gear. 6. Inspect and maintain all HVAC systems and keep a written log. 7. Keep all HVAC systems operating during occupancy. 8. Monitor known IAQ contaminants. 9. Develop a plan to address any needed HVAC system modifications.
Containment Control	10. Develop a plan for noxious and hazardous material control.
Renovation and Remodeling	11. Set a plan for construction and implementation development. 12. Develop an employee awareness program. 13. Keep employees informed of the schedule and its progress.
Employee Information and Education	14. Develop and maintain training materials. 15. Provide initial employee training. 16. Perform follow-up and refresher training.
Record Keeping	17. Develop complaint forms. 18. Develop complaint investigation and resolution forms. 19. Develop HVAC system inspection and maintenance logs.

Table 1: The Proposed OSHA Regulations on IAQ

protracted comment period has delayed OSHA at least a year to move this proposal to a regulation. Table 1, page xiv, summarizes the key elements of these regulations.

As with any new proposed regulation, there has been considerable discussion and anxiety regarding the impact that they will have on business. Despite this, many organizations have already been performing some of the program steps and tasks that would be required and already meet the intent, if not the letter, of the proposed regulation. In addition to OSHA, several professional societies have developed positions or standards relating to IAQ, each addressing some elements and issues of a comprehensive IAQ management plan. Those that pertain to the health care environment in particular include the following:

■ Employee health and morale;

Organization	Standard	Primary Responsibility (Secondary Responsibility)
OSHA	Proposed IAQ Regulation	Engineering (Safety, Risk Management)
NIOSH	Protecting the Health Care Worker	Safety, Infection Control (Engineering, Construction Management)
Joint Commission on Accreditation of Healthcare Organizations	Not directly addressed; issues would fall under Utility and Infection Control standards	Infection Control, Engineering (Safety, Risk Management)
ASHRAE Standards	62-89R	Engineering, Construction Management (Safety)
ASHRAE Research Paper	804-RP	Construction Management, Engineering (Safety, Infection Control)
AIHA	Stock #227-RC-96	Safety, Infection Control (Risk Management)
AIA	96-97 Guidelines for Design and Construction of Hospitals and Health Care Facilities	Planning, Construction Management (Engineering)

Table 2: *IAQ Positions of professional organizations*

- Patient welfare;

- Infectious contamination containment;

- Occupational allergies;

- Transient odors and irritants; and

- Random unpredictable episodic occurrences.

The IAQ positions and standards of these organizations are summarized in Table 2, page xv.

The professional societies of other nations are also developing positions and standards on IAQ, such as the International Society of Indoor Air Quality (ISIAQ), headquartered in Ottawa, Canada.

Clearly, the health care industry as we have known it is undergoing metamorphosis. Economic forces have dictated greater access to high-quality health care at a lower cost, placing greater burdens and responsibilities on health care workers and hospital support staffs. Changing technologies and philosophies regarding patient occupancy have added to that burden, in the forms of exposures to familiar and emerging contaminant sources. While the focus of health care has always been on the patient—and rightly so—it is time to broaden that focus to include the employees and care givers of the organization.

Changes also must occur in terms of recognizing how updated and new heating, ventilating and air conditioning (HVAC) systems will require additional attention with regard to design, operation, maintenance and flexibility. Changes and improvements in medical technology and treatments will also dictate improvements in HVAC system operation, maintenance and reliability. Change is not without cost; but if approached in a positive manner, these costs will generate economic returns.

The key to successfully integrating necessary changes is to focus on the positive including:

- Increased productivity of employees;

- Improved employee relations;

- Decreased absenteeism;

- Decreased workers' compensation claims;

- Decreased union grievances;

- More harmonious working relations, improved energy and operational efficiency;

- Decreased energy consumption;

- Decreased need for additional capacity;

- Increased equipment life as a result of more efficient operation;

- Improved patient satisfaction;

- Increased market share;

- Decreased incidence of nosocomial infections; and

- Decreased potential for litigation.

The information given in this book is meant to be a guide—*your guide.* Use the information presented in Parts I and II as the basis for forming your own opinions as to what is important for your facility. The forms and checklists presented in Part III should be the starting point for developing a program that fits into your operational structure. Maintain your flexibility in designing your program. Focus on the *intent* of what is needed, not on a line item appraisal.

Maintaining a positive outlook is easy to say and difficult to do, but in the complex world of managing the health care environment it is an absolute must. For organizations that have already addressed or begun to address these IAQ issues, the most difficult element of change is accepting the need to change. The case studies shown in Chapter 9 and other organizations that have implemented IAQ programs bear this out.

Finally, we wish to thank all those who contributed to this book. Their dedication and expertise have made this volume possible. In addition, we are grateful to our reviewers. Lloyd Duplechan, HEM, CHMM, is Director of Environmental Risk Management, Kaiser Permanente Regional Environmental Risk Management, Pasadena, California. Douglas Koehler is Director of Facilities Operations, Engineering Department, Hoag Memorial Hospital Presbyterian, Newport Beach, California. Janet Parodi is an executive consultant specializing in managed health care solutions. She has been named to the board of directors of Molina Medical Centers and is currently collaborating on projects with Medical Cities, Dallas, and METIS Technologies Inc, San Francisco Bay

Area. We wish to thank Jeanette Berczi of TransCom Management for coordinating and assisting in the book development process.

Reference

1. OSHA: 29 CFR Parts 1910, 1915, 1926 and 1928 Indoor air quality: Proposed rule. *Federal Register* 59(65): 15968–6309, Apr 4, 1994.

Part I:
Identifying
the Issues

Chapter 1: Sick Building Syndrome: Fact, Fiction, or Facility?

H.E. Barney Burroughs, CIAQP

Mr Burroughs is Director and Principal, Environmental Design International, Ltd, Alpharetta, Georgia. He is a consultant with more than 35 years of experience in the field of mechanical system design, filtration and air cleaning, and indoor air quality analysis. He is a presidential member and fellow of ASHRAE. He is also the chairman of the IAQ certification Board of the Association of Energy Engineers.

The Current Perception of Sick Building Syndrome

A building can be considered sick when it is not functioning properly, which may cause its occupants to be subjected to elements in the indoor environment that can adversely affect their health or their perception of well-being. The conditions and influences that can contribute to this real or perceived poor indoor air quality (IAQ) are varied and complex, as are the individual health effects from poor IAQ. The symptoms of sick building syndrome (SBS) are vague, involving nonspecific upper respiratory discomfort, such as chest tightness, sore throat, coughing, runny nose, burning or itching eyes, headache, lethargy, and fatigue. These symptoms are common to many human maladies, and are often assumed to be the flu or "an improper work ethic." If they actually are SBS symptoms, they will persist for an extended period of time.

The very nature of SBS is that the symptoms will abate or disappear when the affected individual leaves the building.* Discomfort, irritation, and attendant deterioration in productivity is brought about by a prolonged exposure to air constituents that occur normally in low or modest

---

* In buildings having higher population density and more consistent occupancy such as long term care facilities, SBS is strongly considered when IAQ-related complaints reach 20% of the building population. However, this rule of thumb may not be a realistic indicator in the health care environment because of increased sensitivity of the occupants.

concentrations. (Later in this chapter, I discuss specific categories of exposure in more detail.) The constituents of the air that are of concern are normal and routinely occurring components, such as the following:

■ Airborne dust and inorganic particulate matter;

■ Airborne volatile organic chemicals and vapors; and

■ Allergens, pollens, environmental mold, and bacteria.

The generic cause of SBS is the amplification of one or more of these constituents to an elevated contaminant concentration, resulting in potentially harmful levels. Small concentrations of harmful or toxic compounds can be tolerated by the human body, and high concentrations of the most benign compound can be very detrimental. It is therefore necessary to gain an understanding of the systems, sources, and practices that promote or facilitate the amplification of environmental constituents to harmful contaminant concentrations in order to diagnose, prevent, and mitigate the causes of poor IAQ.

The ecology of the health care facility: A constituent/exposure model. It may be helpful to think of the health care facility as a habitat consisting of the site, the building, and the air within the health care facility. This habitat is in an ecosystem defined as the interrelationship among the facility occupants and its components. Is the habitat free of harmful constituents and odors, comfortable, and "fresh"? Is the air path clean and dry, with all systems functioning as intended? Is the building sound, dry, and leak free and under positive pressure with respect to outdoor ambient pressure? Is the site close to contaminant sources? These are all questions that must be dealt with for the model to be accurate and successful.

Time is a key element of this model because situations, systems, building usage and loads, and performance will all change. Aging mechanical systems deteriorate, break, and fail; this makes maintenance and operating practices another necessary component to our model.

The occupants are the most disruptive, albeit essential, elements of the model. While there are always special comfort needs and hygiene concerns for the patients, the broader issue in this book is the environment in which the health care worker performs his or her duties. Disease transmission, laboratory related activities, sterilization, and other support activities, for example, provide a level of potential contamination unique to the health care facility. The facility management team must understand each component of the ecosystem and the SBS causative and influencing factors, and how to use them as part of the SBS problem solution.

The heating, ventilation, and air conditioning system: A "pathway for pathogens."[1] The air pathway is an essential ingredient to understanding environmental quality in the health care facility. The purpose of the heating, ventilation, and air conditioning (HVAC) system is to maintain acceptable comfort conditions and to remove contaminated air from the building, which makes it a key element in the air pathway. Comprehending IAQ-related complaints requires an understanding of the HVAC system.

Outdoor air enters the HVAC system, where it is cleansed by filtration systems to remove particulate matter before it enters the air distribution system. The HVAC system then conditions the air to appropriate temperature and humidity levels and delivers it to each occupied zone in the building. A failure at any step could subject the occupants to potential discomfort and exposure to contaminants, possibly resulting in instances of IAQ complaints. After the air enters the conditioned space, it is withdrawn through the return duct system and either returned to the HVAC unit or exhausted to the atmosphere. A failure here permits air stagnation and may promote the concentration of airborne contaminants and building pressurization problems, encouraging unfiltered air to reach occupants, again causing potential discomfort.

The results of constituent amplification. There are a number of factors that can contribute to the HVAC system being a contaminant amplifier or source, such as the following:

■ Basic design flaws;

■ System performance and mechanical failures;

■ Control system flaws, both in design and operation;

■ Dirt and contaminant buildup within the HVAC air distribution system;

■ Pooling water within the HVAC unit; and

■ Negative pressurization.

A malfunctioning HVAC system will fail to deter amplification at other sites in the facility ecology. For example, improper location of the outdoor air louver at a loading dock can cause workers to ingest diesel fumes. Insufficient air delivery to the occupied space may result in the HVAC system's failure to adequately dilute odors, vapors, or viable pathogens below harmful levels. Negative pressurization will cause envelope leakage and potential fungal growth within the interstitial space of the envelope. Poor humidity control will also contribute to bioaerosol growth throughout the space.[2] Inadequate filtration will create a habitat in the distribution system favorable to fungal growth. Improper air balance will allow contaminated air to flow into critical patient "clean" areas or into an exposure pathway for health care workers. Improper exhaust location could allow cross contamination.

The net effect of any of these system flaws is the unacceptable presence of contaminants in the indoor environment. Contaminants can be introduced into the indoor environment by patients or the activities of the health care staff in executing their treatment. The result of the exposure to these contaminants is highly dependent upon the dosage exposure and the potency of the contaminant. The impact of low-level exposure is usually discomfort: stuffiness, hot or cold calls, and so on. Higher levels of exposure will elevate discomfort to the irritation level: itching, burning eyes; breathing difficulties; and so on. As the exposure level increases, so does the complaint level, reaching the area of mild illness symptoms. Serious exposures to toxic and/or pathogenic constituents can bring about sensitization, illness, or permanent disability, and may progress to building-related illness (BRI) (see Chapter 2).

Any of these levels of exposure can result in diminished productivity of the health care staff. This can take the form of heightened fatigue and loss of stamina, lack of focus and concentration, mistakes, irritability, absenteeism, and worker's compensation claims (see Chapters 5 and 8).

A more subtle but disastrous consequence is adverse publicity generated by a negative image of the health care facility as perceived by the general public.

The Development of SBS in Commercial, Institutional, and Health Care Facilities

Energy and ventilation. Since the mid-1970s, concerns for energy conservation management have dominated building design and operation objectives. As a result, building envelopes have been better insulated and constructed more tightly to avoid infiltration and heat loss. Mechanical systems (excluding 100% outside air systems) were specifically targeted by reducing outdoor ventilation air. In the name of energy conservation, facilities severely restricted the outdoor ventilation air, resulting in a rash of complaints and adverse building-related health effects. This was the birth of SBS. Early solutions to these problems assumed that increased ventilation would completely solve the complaint. Unfortunately, this assumption falsely focused on ventilation rates as both the *cause* and the *cure* for all SBS effects.

Factors contributing to the evolution of IAQ and SBS problems. There are a number of other factors that contributed to the emergence of IAQ issues:

■ *Regulatory focus* and prevailing national priorities brought public attention and awareness to environmental issues such as radon, asbestos, fertilizers, pesticides, and other hazardous materials. Handling, shipping, storing, and content reporting and labeling requirements were imposed on products having chemical content.

■ *Litigation.* A number of IAQ cases with large settlement costs have been widely publicized. Most informed building occupants are aware of (and perhaps unduly apprehensive about) building-related health risks. Legal precedents have already been established that declare the building to be a product for which "strict liability" law applies. Related rulings have declared the building owner to be responsible for the quality of the indoor environment and fully liable for negligence in operation and maintenance.

- *Budget constraints.* Health care facilities face the double-headed ax of reducing operating costs and increasing the quality of health care. Value engineering (a euphemism for cost cutting) has eliminated, restricted, or compromised components and systems in the construction process. Because commissioning and project management has been similarly eliminated, problems of system inadequacies are discovered after the fact. Life-cycle costing usually loses out to the first cost decisions in this process, often leading to SBS problems down the road. The maintenance budgets for health care facilities have also faced the ax. In fact, deferred maintenance has been implicated in most of the reported SBS incidents. (This is a theme that will be repeated often throughout this book.)

- *Changes in building components and furnishings.* There is a greater prevalence of petrochemical-based synthetics for all kinds of building and furnishing components. These products have created a chemical "stew" of volatile chemical compounds (VOCs) that out gas heavily when new, but also yield contaminant load continuously over the lifetime of the products. Modern furnishings and equipment yield their own burden of contaminants, such as noxious odors, respirable dust, and ozone.

- *Aging building components* contribute to IAQ. Systems wear, their performance degrades, and they become obsolete. This can be debilitating to health care facilities that have constant levels of demand and expectation from both staff and equipment. Sound proactive maintenance programs are necessary to avoid unscheduled interruptions that contribute to the gradual deterioration of environmental quality.

- *Improper, inadequate, or untimely response to IAQ* complaints. Ignoring SBS complaints or postponing correction of deterioration of building system performance can escalate minor complaints into major liabilities. Indifferent handling of SBS complaints from health care staff members can alienate workers and create labor relations barriers. (Examples of SBS complaint and interview forms in Part III can help prevent molehills from becoming mountains.)

Causes of Poor IAQ and SBS in Health Care Environments

The causes of poor IAQ are difficult to pinpoint because they are complex, involving sources, systems, operations, and occupant behavior. There is seldom a *single* cause for IAQ-related health effects, but a majority of the causes are related to the facility system and operational factors,[3] including the following:

- *Low ventilation rates and poor ventilation efficiency or inadequate delivery.* Extended care facilities, medical office complexes, outpatient areas, and remote outreach facilities are susceptible and are potential risk areas for insufficient ventilation levels. Current ventilation guidelines base ventilation rates on volume per minute per occupant. The assumption is that the occupant activities are responsible for the dominance of contaminants in the conditioned space.

  Most systems fix the ventilation rate at the expected (or design) occupancy. If building use changes, occupancy might vary, which would alter the ventilation requirements. High levels of nonoccupancy contaminants, such as those coming from purposeful housekeeping compounds or specific health care procedures, can affect the ventilation system. System component failure can decrease the level of the ventilated air.

  Poor ventilation efficiency is the failure of ventilation air to be delivered in proper volume to the occupied zone. Malfunctioning control systems can restrict air. Partitions, screens, and medical equipment can create barriers to air flow, as can pressure imbalances between zones within the building.

- *Contaminated outside air.* Health care facilities use a great deal of outside air, which makes them vulnerable to exterior contaminants. Health care facilities usually use high efficiency particulate arrestance (HEPA) filters, but high particulate loads cause premature filter loading. Gas phase filtration is required to address fumes, vapors, and gases, and very few health care institutions have this capability.

◼ *Poor or inadequate filtration selection or performance.* While this is not an issue for hospitals, satellite facilities and medical office buildings may have low-efficiency filters as artifacts of design or prior operation. Poor filtration can lead to fouling of coils and contamination of ductwork. Filters that are not properly monitored and serviced can become ill-fitted in their brackets and thus become a contaminant source due to fungal growth.

◼ *Water incursion, including poor humidity control.* Roof leaks, condensation, high humidity, and moisture carryover from cooling coils are examples of adverse effects of water on IAQ. Chronic leaks that sustain fungal growth anywhere in the facility are a particular problem. The control of humidity is critical in the avoidance of fungal growth, which can happen almost overnight.

◼ *Improper air balance and pressure relationships.* Balance and proper pressure differentials are critical. Positive pressure in a building is the best method of minimizing infiltration of untreated outside air into the building environment. Correct air flow direction, velocity, and proper pressure differential relationships are essential to ensure pathogen control in critical care areas.

◼ *Inadequate or deferred maintenance practices.* As stated earlier, and will oft be repeated, a sound proactive preventive maintenance program is a facility's first line of defense against IAQ-related issues. This is going to require a better understanding by hospital administration, which places a burden of education on the facility's engineering department.

◼ *Purposeful activities with negative impact,* such as the use of certain housekeeping chemicals and pest control compounds. This is a cooperative educational area and is covered in more detail in Chapter 6.

◼ *Poor human resource and labor relations* as the result of poor communication and slow response to complaints. This is covered in detail in Chapters 5 and 6.

◼ *Related factors such as noise, lighting, ergonomics, and stress* also add to the perception that all is not well with the air in the building.

Constituents of Concern to Health Care Workers

Health care facilities are veritable repositories of contaminants, which are usually under adequate control. These contaminants include VOCs such as gluteraldehyde, xylene, and formaldehyde. (These compounds are discussed in more detail in later chapters.)

Other gaseous compounds can be introduced into the health care facility, such as sulfur dioxide or hydrogen sulfide, traceable to leaks in the plumbing vent system or from outside sources. Ozone can come from general external pollution and from electrical equipment within the hospital. One source of internally produced ozone is from printers, copiers, and facsimile machines that use a laser engine. The laser process generates ozone, which is normally broken down by the ozone filter within the unit. If this filter fails, free ozone is emitted into the area.

Odors are a common thread to all contaminants, but especially to VOCs. For the most part, odors are just that: an unpleasant smell. Unfortunately, the complaints that follow an odor incident are passionate and highly emotional and tend to be a nightmare for facility engineers.

Carbon dioxide is both naturally occurring and a byproduct of the human metabolism. Carbon dioxide at the normal concentrations found in buildings is not a concern; it does represent an effective surrogate for other body bioeffluents as well as other gaseous compounds within the space. In the past,[4] levels under 1,000 PPM were considered acceptable. ASHRAE's proposed IAQ standard 62R may change that philosophy; a more acceptable model may target the inside level of carbon dioxide at no more than 700 PPM above the ambient concentration.[5]

Particulate matter is present in health care facilities and comes from a wide array of sources. Particles are generated indoors from paper handling activities (medical records is a high source generator of cellulose particles); surgical scrubs, gowns, and linens (large amounts of fibers); and people (we shed dead skin cells continuously). Even walking on carpeted floors, and sitting in upholstered chairs generates dust. Respirable dust particles, those that can reach the deep lung cavities and bronchial tubes, are the ones of great concern.

Tactics for the Control and Removal of Contaminants

There are three possible tactics for the control and removal of airborne contaminants:

- *Source control.* This is the most efficient, and can be the most cost-effective of the control tactics. For example, it may be most cost-effective to relocate the makeup air louver away from the source of vehicle exhaust. However, posting signs prohibiting the idling of vehicle engines while in the dock area is an example of source control through management. A thorough review of building components, furnishings, decorations, equipment, management practices, housekeeping techniques and supplies, and purposeful chemicals will identify sources of airborne particles and gases. These sources can be eliminated or reduced by substitution, isolated from potential exposure to occupants by either distance or timing, or exhausted from the space before reaching the public air stream.

- *Filtration and air cleaning* can effectively eliminate very high levels of particulate contamination. Composite filter banks with a final filter efficiency of at least 95% removal of particles of 1 micron size is the critical quality target. The final filter system is usually located on the positive side of the air handler. If gas phase filtration is installed, it is located prior to the final filter system. A HEPA-type (99+ percent at 0.3 micron size particle) terminal filtration could be installed in very critical areas such as operating rooms and critical care areas, with laminar flow HEPA installations at the operating table or other critical care sites.

  Self-contained, fan-powered filtration units can be used to provide localized filtration control for areas such as patient isolation rooms or triage.

- *Dilution.* Ventilation has been the classic method of contaminant control and reduction. This classical technique is successful in general contaminant control, but it has liabilities:

- Outdoor air is no longer considered to be "fresh air" and may be contaminated.

- The temperature and humidity of outside air create a load on the HVAC system making it an energy liability.

- Using outside air is less efficient than employing source control or filtration. In most facilities, only the critical care areas use 100% outside air. Other areas employing fractional ventilation must rely on diffusion and thorough mixing of the supply air for the dilution process to be effective.

These tactics can be used together to reduce the general load of contaminants introduced into the conditioned space. This will make for better IAQ and should reduce instances of SBS.

References

1.   Burroughs HE: The role of risk assessment as a mitigation protocol for IAQ causes. World Energy Engineering and Environmental Congress, 1991, proceedings.

2.   ASHRAE Standard 55: Thermal engineering. American Society of Heating, Refrigerating, and Air Conditioning Engineers, Inc, 1994.

3.   Turner W: Indoor air quality. In Burroughs HE (ed): *Energy Management Handbook*. Lilburn, GA: The Fairmont Press, 1996, p 484.

4.   ASHRAE Standard 62: Ventilation for acceptable indoor air quality. American Society of Heating, Refrigerating, and Air Conditioning Engineers, Inc.

5.   ASHRAE Standard 62R Draft: Ventilation for acceptable indoor air quality. American Society of Heating, Refrigerating, and Air Conditioning Engineers, Inc.

Chapter 2: Building-Related Illness with a Focus on Fungal Issues

Philip R. Morey, PhD, CIH

Dr Morey is Vice President and Director of Microbiology, AQS Services, Inc, Gettysburg, Pennsylvania. Dr Morey has been actively involved in the industrial hygiene community and in the bioaerosol and indoor air quality fields for more than 30 years. He is a member of the International Academy of Indoor Air Sciences and Co-chair of the ISIAQ task force on control of moisture problems affecting biological IAQ.

Building-related illness (BRI), as distinguished from sick building syndrome (see Chapter 1), in this chapter describes clinically diagnosed health problems caused by exposure to contaminants in the building. Nosocomial Legionnaires' Disease and aspergillosis are examples of BRI. Cancer and other diseases caused by exposure indoors to sensitizing and allergenic agents such as airborne latex and other contaminants (see Chapter 3), are potentially BRIs.

SBS and BRI can occur simultaneously in the same building; but whereas the annoyance and discomfort symptoms characteristic of SBS remit quickly after cessation of exposure, BRIs can result in severe morbidity and mortality.

BRIs are often associated with poor maintenance and housekeeping practices, which can lead to the growth and/or accumulation of environmental microorganisms, especially fungi, in buildings. In this chapter I summarize the conditions that precipitate this growth and offer suggestions for how to prevent it.

Culturable Fungi and Other Microorganisms in Indoor Environments

Several types of microorganisms are the cause of BRIs in health facilities.

Fungi. In general, the kinds of fungi found in buildings without moisture problems are similar to those found outdoors.[1,2,3] A sure sign of a moisture problem indoors occurs when the diversity of fungi indoors compared with those outdoors is out of balance. Table 2–1, below, compares the diversity of fungi in three hypothetical buildings of a medical facility: Building #1 without moisture problems, Building #2 with dampness in one room, and Building #3 with severe flooding. *Cladosporium* spp. dominates the fungi in Building #1 as well as in the air outdoors; *Penicillium* spp. dominates the fungi in Building #2; and *Aspergillus glaucus,* a species not present in the outdoor air at the time of sampling, is the major species in Building #3. Note that the concentration of this *Aspergillus* spp. approached 10^5 colony-forming units per cubic meter of

Description of Building and Sampling	Concentration (cfu/m³)	Predominant Kinds of Fungi	
Building #1, absence of moisture problems, quiescent sampling, N=5	220	64% 17% 15%	*Cladosporium* spp. *Penicillium* spp. non-sporulating
Building #2, damp carpet in one room, quiescent sampling, N=5	470	57% 31% 7% 4%	*Penicillium* spp. *Cladosporium* spp. *Aspergillus versicolor* non-sporulating
Building #3, all carpets had been flooded, some carpet still damp, quiescent sampling, N=4	2,500	70% 15% 8% 6%	*Aspergillus glaucous* (*Eurotium* spp.) *Penicillium* spp. *Cladosporium* spp. non-sporulating
Building #3, sampling while walking on carpet, N=2	10^5	80% 20%	*Aspergillus glaucous* (*Eurotium* spp.) *Penicillium* spp.
Outdoor air around buildings, N=7	170	67% 21% 10%	*Cladosporium* non-sporulating *Penicillium* spp.

Malt extract agar; Volumetric impactor operating at 0.18 m³/min; 25 – 28°C incubation.

Table 2–1. *Airborne Fungi Present in Buildings with and without Moisture Problems*

air (cfu/m³) when the damp carpet in Building #3 was disturbed by walking. Table 2–1 shows that chronically damp porous materials provide niches for the growth of fungi, demonstrating that dampness and flooding should be avoided in all areas in medical facilities.

Table 2–2, below, provides data on thermotolerant (35–37°C) fungi found in and around a hypothetical medical center. *Aspergillus* and *Penicillium* species were present outdoors. Although the total concentration of thermotolerant fungi was reduced in organ transplant patient area #1, *Aspergillus fumigatus* was detectable at the nursing station and

Description of Sampling Location	Concentration (cfu/m³)	Kinds of Fungi
Outdoor air at grade near main entrance; absence of major construction activities; N=19	19 5 4 4 1	Total fungi (all taxa) *Aspergillus niger* *Aspergillus fumigatus* *Penicillium* spp. *Paecilomyces* spp.
Nursing station #1 serving organ transplant patients; carpeted; N=15	13 1.5	Total fungi (all taxa) *Aspergillus fumigatus*
Patient rooms, zone #1; non-carpeted; N=11	4 2 1	Total fungi (all taxa) *Aspergillus fumigatus* *Aspergillus niger*
Nursing station #2 serving organ transplant patients and adjacent corridors; carpeted; N=12	10 9	Total fungi (all taxa) *Aspergillus fumigatus*
Patient rooms, zone #2; non-carpeted; N=4	8 7	Total fungi (all taxa) *Aspergillus fumigatus*
Peak concentrations in outdoor air at grade level and HVAC inlet when major construction project occurs (site excavation)	20,000 10,000 700 500 25 25	Total fungi (all taxa) *Penicillium* spp. *Aspergillus niger* *Aspergillus fumigatus* *Aspergillus terreus* *Aspergillus flavus*

Malt extract agar; Volumetric impactor operating at 0.18 m³/min; 37°C incubation.

Table 2–2. *Airborne Thermotolerant Fungi in and Around a Medical Facility*

in patient rooms. In organ transplant patient area #2, *Aspergillus fumigatus* accounted for at least 80% of the thermotolerant fungi. This air sampling data indicated that the indoor environment of the facility is mycologically unacceptable and that the HVAC system filters are ineffective and/or that there are strong amplification niches for fungal growth in the heating, ventilating, and air conditioning (HVAC) system or in the interior spaces.

Construction activities (see Chapters 4 and 7) are a common occurrence around many medical facilities, and soil excavation results in aerosolization of fungi. Table 2–2, shows that concentrations of thermotolerant fungi increased above background outdoor levels when prevailing winds carried soil particles toward the medical facility and its HVAC system outdoor air inlets.

For organ transplant and other at-risk patient areas, numerical guidelines for thermotolerant fungi may be appropriate. Arnow et al.[4] found that epidemic aspergillosis occurred when the level of 37°C *Aspergillus fumigatus* and *Aspergillus flavus* was 1 cfu/m^3 or higher. A facility can minimize building-related aspergillosis when it provides mycologically high-quality air by controlling moisture and dust and by practicing good HVAC system maintenance most notably by including terminal HEPA filtration in the system.

Other microorganisms. BRIs in medical centers can be caused by microorganisms other than fungi. Reservoirs for a wide variety of microorganisms, including yeasts and gram negative bacteria can occur on wet surfaces such as drain pans and cooling coils and in water sumps serving cold water humidifiers. Again, these microorganisms, indicated by the presence of a biofilm on a wet surface, are largely due to poor maintenance.

Legionella. There are more than 30 different *Legionella* species, but the one of most interest is *Legionella pneumophila.* This bacterium lives in bodies of water rivers, lakes, streams, and so on. One can find it in the water supply of health care facilities with no identified disease. Domestic and mechanical water systems that are maintained at 35°-45°C provide optimal conditions for the growth of *Legionella,* which can cause

Legionnaires' Disease in susceptible patients. Pontiac Fever is a less serious form of legionellosis with flu-like symptoms. Again, as might be suspected, most outbreaks of legionellosis are the result of poor water supply maintenance.

Legionella species are not identified on routine cultures; therefore, the disease is thought to be grossly underreported. Legionellosis symptoms can range from a mild cough with a low grade fever, to a rapidly progressive pneumonia. It can disseminate through the blood stream to other organs resulting in multi-organ failure and death. For this reason, one case of nosocomial legionellosis is sufficient cause to investigate the water supply and airflow in a medical facility. Outbreaks can occur following activities that disturb and dislodge the scale in water pipes, such as repairs or modifications to the domestic plumbing system. Stagnation of water, such as dead-end piping, can result in amplification of *Legionella*. Principles for controlling *Legionella* are well-known[5,6,7] and a routine testing program for culturable *Legionella* is useful in verifying the effectiveness of preventive maintenance programs.

Fungal pathogens. The fungal pathogens *Histoplasma capsulatum* and *Cryptococcus neoformans* are associated with bird and bat fecal material and can cause infection. These pathogens are difficult to detect by air sampling methods[3]; therefore, the visual presence of bird or bat feces in buildings or in the HVAC systems should be viewed as a positive presence of these pathogens. Maintenance personnel should remove all fecal material using procedures that prevent the dissemination of spores into the occupied areas of the facility.

Sources of Fungi in a Medical Facility

Sources of fungi include the outdoor air, construction dust, wet niches in the HVAC system, and living plants. Providing highly filtered outdoor air to patient areas has historically resulted in a decline of aspergillosis.[8]

Instances of aspergillosis during construction, soil excavation, and building demolition are often associated with aerosolization of fungi[9] (see Table 2–2, page 17) and entry of spores into a building. Dirt and dust that occur on plenum surfaces in the HVAC system can be assumed to contain thermotolerant *Aspergillus fumigatus.*[10]

Dust containing spores can settle on dry surfaces of the HVAC system and remain as a reservoir until disturbed. Fungi can grow on air stream surfaces in HVAC systems if adequate moisture is present. The culturable *Aspergillus fumigatus* detected in a hypothetical medical facility described in Table 2–3, below, was caused by moisture in air handling units. Niches for fungal growth include chronically damp internal lining in the HVAC system.[10,11] Facilities engineers can correct this kind of existing contamination by physically removing the insulation.

Description of Source of Fungal Contamination	Concentration and Kinds of Fungi
Airstream surfaces in HVAC system air handling units including dusts on metal surfaces, exposed insulation, and filters	35 of 48 samples contain culturable fungi; maximum concentration 2×10^4 cfu/square inch; 26 of 35 positive samples contain *Aspergillus fumigatus*
Wet final filters in air handling units (above)	4 of 6 samples contain *Aspergillus fumigatus;* peak concentration 104/square inch
Insulation in ductwork serving transplant patient room	Airstream surface of insulation contains *Aspergillus fumigatus* at concentrations of 10^2 to 10^4 cfu/square inch
Dusts collected from upper surface of ceiling tiles above corridors and patient rooms	Concentrations of fungi range from 10^1 to 10^4 cfu/gram. Dominating fungi vary in samples and include *Aspergillus fumigatus, A. niger, A. glaucus,* Rhizopus, and yeasts
Dusts collected from patient rooms and zones undergoing renovation	Concentrations of fungi as high as 10^6/gram. Dominating taxa in dust samples varies and includes *Cladosporium spp., Penicillium spp., Aspergillus versicolor, A. niger,* and yeasts.

Samples analyzed by serial dilution at 37°C on Malt extract agar.

Table 2-3. *Thermotolerant Fungi Present in HVAC System and Patient Areas in a Medical Facility*

Porous materials, such as ceiling tiles, carpet, upholstery, and drapes, can provide niches where fungal spores can settle and accumulate. Settled dust on the upper ceiling surface above patient areas can contain a variety of fungi (Table 2-3, page 20). Dust removal must be performed with caution; vacuum cleaning is associated with increases in airborne culturable fungi.[12]

Moisture entering wall board by capillary action from wet carpet can cause fungal growth. Because massive numbers of conidia (spores) occur on visually contaminated (moldy) surfaces, facilities engineers should consider the air in rooms with visually moldy materials to be heavily contaminated irrespective of HVAC supply air filtration.[13]

Prevention of Microbial Growth

The best way to prevent fungi growth is to limit the availability of moisture in building infrastructure and on air stream surfaces in the HVAC system. (I use the terms *water activity* [a_w] and *equilibrium relative humidity* [ERH] to describe the availability of moisture for fungal growth.[14]) Fungi growth on susceptible surfaces appears to occur when the a_w is 0.80 or the surface ERH is consistently maintained at 80% or more for several days.[15,16] When the a_w of a material is less than 0.65 (surface ERH does not exceed 65%), virtually no fungus will grow, even under optimal nutrient and temperature conditions.[2]

The kind of fungus that grows on a surface is determined by moisture availability (a_w) as well as by substrate and temperature. Different kinds of fungi can grow in the same construction material, depending on the availability of moisture. Therefore, the presence of "signature flood fungi" such as *Fusarium* and *Chaetomium* indicate chronic flooding conditions. Fungi such as *Aspergillus versicolor* and *Eurotium* species suggest damp but not wet conditions.

All medical facilities should have a plan of action for moisture removal. Table 2–4, page 22, illustrates the kinds of microbial contamination that can occur on building finishes if moisture is not removed promptly after a flood. Fungi can visually contaminate wet wall board and carpet and be dispersed into the indoor air.

Description of Source of Fungal Contamination	Concentration and Kinds of Fungi
Wall board that was soaked for several weeks and became visibly moldy on occupied side of wall.	Total concentration of 25 – 28°C fungi 5 X 10⁶/square inch; 80% *Stachybotrys chartarum;* 20% *Aspergillus* spp. Including *A. versicolor, A. ustus,* and *A. nidulans*
Wall cavity surface of same wall board as in previous example (soaked for several weeks)	Total concentration of 25 – 28°C fungi 7 X 10⁶ cfu/square inch; 89% *Fusarium* spp., 10% *Ulocladium* spp., and 1% *Penicillium brevicompactum*
Carpet that remained wet for several weeks	Total concentration of 37°C fungi was 10⁵/gram; *Aspergillus niger, A. fumigatus,* and *Paecilomyces* spp. were major isolates
Air in rooms with soaked carpet; N=4	Total concentration of 37°C fungi was 1,550 cfu/m³; *Aspergillus niger* and *Paecilomyces* spp. were predominant isolates in all samples

Table 2–4. *Fungi Present in Interior Finishing Materials and in Indoor Air After a Major Flood*

Fungi can grow in any component of the building where moisture is not limited, such as after a flood. But engineers must also be aware of more subtle moisture problems that can occur in building envelopes in the heating and cooling seasons.[2] In cold seasons, the interior surface of exterior walls can become sufficiently cool (because of the thermal gradient across the wall) that the ERH of room air approaches or reaches 100% at the wall surface. The resulting high moisture content of the wall board or plaster surface creates an environment where various fungi can grow. In cooling seasons, infiltrating humid air encountering cool finishes on the interior surface of exterior walls can spur fungal growth. Fungal contamination in envelope walls in hot humid climates is greatest in those buildings that are overcooled and under negative pressure.[10,15]

In general, engineers can control microbial growth in or near HVAC equipment by doing the following:

■ Keeping the air stream surface dry;

■ Making the air stream surface cleanable or replaceable; and

■ Using a material that is not susceptible to biodeterioration.[15]

HVAC systems should be designed and operated to prevent airstream surfaces downstream of cooling coils from becoming wet.[17] Avoid water stagnation in drain pans and allow condensate to drain freely.

Control of Existing Fungal Contamination and Renovation Dusts

There are several well-described procedures for removing visible fungi.[1,2,18,19] Those described in consensus documents, though not stringent enough for medical facilities, do provide a starting point for remediation protocols. In health care facilities, failure to contain renovation dusts can result in BRIs such as aspergillosis.[20,21,22] Facilities engineers should assume that dusts from wall cavities, ceiling systems, and carpet contain culturable fungi including *Aspergillus* species. Containment barriers must be both rigid and impervious.[9,21] Barriers must extend from floor slab to ceiling slab, all utility penetrations must be sealed, and renovation work areas must be constantly negatively pressurized. Administrative controls are required to make certain that renovation dusts are not tracked into other areas, particularly patient care areas. BRIs associated with poor maintenance can thus be kept to a minimum.

References

1. Health Canada: *Fungal Contamination in Public Buildings: A Guide to Recognition and Management.* Ottawa: Federal-Provincial Committee on Environmental and Occupational Health, 1995.

2. ISIAQ: *Guidelines for Control of Moisture Problems Affecting Biological Indoor Air Quality.* Ottawa: Task Force 1, International Society of Indoor Air Quality, 1996.

3. AIHA: *Field Guide for the Determination of Biological Contaminants in Environmental Samples.* Fairfax, VA: Biosafety

Committee, American Industrial Hygiene Association, 1996.

4. Arnow P, Sadigh M, Costas C, et al: Endemic and epidemic aspergillosis associated with in-hospital replication of *aspergillus* organisms. *J Inf Dis* 164:998–1002, 1991.

5. ASTM: *Guide for Inspecting Water Systems for Legionella and Investigating Possible Outbreaks of Legionellosis (Legionnaires' Disease and Pontiac Fever)*. West Conshohocken, PA: American Society for Testing and Materials, 1996.

6. HSE: *The Control of Legionellosis Including Legionnaires' Disease*. Health and Safety Series Booklet HS (G) 70. Sudbury, UK: Health and Safety Executive, 1994.

7. Freije M: *Legionellae Control in Health Care Facilities*. Indianapolis: HC Information Resources, Inc, 1996.

8. Rose H, Hirsch S: Filtering hospital air decreases *aspergillus* spore counts. *Amer Rev Resp Dis* 119:511–13, 1979.

9. Streifel A: Aspergillosis and construction. In R. Kundsin, ed: *Architectural Design and Indoor Microbial Pollution*. New York: Oxford University Press, 198–217, 1988.

10. Morey P: Control of indoor air pollution. In P Harber, M Schenker, J Balmes, eds: *Occupational and Environmental Respiratory Medicine*. St. Louis: Mosby, 981–1003, 1995.

11. Fox B, Chamberlin L, Kulich, et al: Heavy contamination of operating room air by *penicillium* species: Identification of the source and attempts at decontamination. *Amer J Inf Cont* 18:300–306, 1990.

12. Hunter C, Grant C, Flannigan B, et al: Mold in buildings: The air spore of domestic dwellings. *Intern Biodeterioration*, 24:81–101, 1988.

13. Streifel A, Stevens P, Rhame F: In-hospital source of airborne *Penicillium* species spores. *J Clin Microb* 25:1–4, 1987.

14. Flannigan B: Approaches to assessment of microbial flora in buildings. In *IAQ '92 Environments for People*. Atlanta: American Society of Heating, Refrigerating, and Air-Conditioning Engineers, Inc, 139–45, 1992.

15. Morey P: Mold growth in buildings: removal and prevention. In *Proc 7th Intern Conf on Indoor Air Quality and Climate,* 2:27–36, 1996.

16. Hunter C, Sanders C: Mold. *Energy and Condensation in Buildings and Community Systems Programme: Final Report,* 1:2.1–2.30; Condensation and Energy Sourcebook. Brussels, Belgium: Intern Energy Agency, Annex 14, Leuven University, 1990.

17. Morey P: Suggested guidance on prevention of microbial contamination for the next revision of ASHRAE standard 62. In *IAQ '94 Engineering Indoor Environments.* Atlanta: American Society of Heating, Refrigerating, and Air-Conditioning Engineers Inc, 139–48, 1994.

18. New York City Dept Health: *Guidelines on Assessment and Remediation of Stachybotrys atra in Indoor Environments.* New York: NYC Human Resources Administration, Mt. Sinai—IWS Occupational Health Clinic Center, 1993.

19. Morey P: Microbiological contamination in buildings. Precautions during remediation activities. In *IAQ '92 Environments for People.* Atlanta: American Society of Heating, Refrigerating, and Air-Conditioning Engineers, Inc, 171–78, 1992.

20. Arnow P, Andersen R, Mainous P, et al: Pulmonary aspergillosis during hospital renovation. *Amer Rev Resp Dis* 118:49–53, 1978.

21. Krasinski K, Holzman R, Hanna B, et al: Nosocomial fungal infection during hospital renovation. *Infection Control* 6:278–82, 1985.

22. Opal S, Asp A, Cannady P, et al: Efficacy of infection control measures during a nosocomial outbreak of dissseminated aspergillosis associated with hospital construction. *J Infectious Diseases* 153:634–37, 1986.

Chapter 3: Risk Factors for Occupational Exposures in Health Care Professionals

John F. McCarthy, ScD, CIH

Dr McCarthy is President, Environmental Health and Engineering, Newton, Massachusetts. He is a recognized expert in the field of indoor air quality, and a leader in identifying emerging issues. He serves on the Indoor Air Quality Engineering Research Subcommittee of the EPA's Science Advisory Board, among other committees, and is widely published in the areas of exposure assessment, control measures, and program development.

The modern hospital facility embodies the goals and aspirations of today's sophisticated health care industry. In these facilities, we care for the most susceptible of our population. We strive to utilize the most sophisticated technology, advanced therapeutic treatments, and intensive personal interaction to improve an individual's well-being. Yet the need to meet increasingly stringent requirements of health care provision places intense pressure on the facilities themselves and the health care workers (HCWs) who occupy them.

During the struggle to fulfill standards and provide excellent care (while keeping costs to a minimum), health care facilities often face safety issues, particularly in relation to indoor air quality (IAQ). In addition to sick building syndrome (discussed in Chapter 1) and building-related illness (discussed in Chapter 2), hospitals by their very nature as treatment and healing facilities present many potential hazards that require constant vigilance to maintain a safe and healthful environment. Control of the environment is complex but absolutely essential due to the following:

■ The presence of pathogenic agents such as hepatitis-B and *mycobacterium tuberculosis (M tuberculosis)*;

■ The need to disinfect and sterilize medical devices, equipment, and surfaces; and

■ Other chemical, physical, or biological agents used for treatment.

In distinct contrast to the dramatic advances in equipment and treatment that distinguish modern health care delivery, the design, operation, and maintenance of most hospital facilities have generally not evolved as rapidly. The need for advanced isolation, increased ventilation effectiveness, and environmental control has stressed many hospitals' existing mechanical systems. Furthermore, the fact that most areas are occupied continuously, requiring systems to operate 24 hours a day, seven days a week, presents challenges to preventive maintenance, repair, and system upgrades.

This chapter presents an overview of the environmental risks present in the hospital setting. The opportunities to improve provision of health care also present new challenges for building performance and environmental control. Maintaining a safe and healthful hospital environment depends on the continuous review of new procedures and the rapid assessment of possible environmental hazards. Through anticipation of potential risks, thoughtful intervention, effective communication, and input about integrating environmental concerns into facility design, we can provide a hospital environment that meets the ideals of health care for all occupants.

Exposure Types

As discussed earlier, certain types of indoor exposures are more prevalent in hospital settings than in other public facilities. The following sections describe some of the common exposures.

Sensitizing and allergenic agents. A number of materials commonly found in hospitals have strong sensitizing and allergenic properties, including histamines, gluteraldehyde, formaldehyde, latex allergen, hexachlorophene, and psyllium laxatives. It is well established that sensitization can occur during the preparation of pharmaceutical products, especially antibiotics. Kern and Frumkin found that the respiratory therapists in their study reported significantly higher asthma rates than did matched control subjects.[1] Although the specific sensitization mechanism has not yet been discovered, these results have been replicated.[2]

Latex allergen has also been shown to elicit various dermatologic and systemic reactions via type I immune mechanisms. (This is discussed in more detail in the "Emerging Hazards" section, page 33.)

Irritants. Several compounds or mixtures of compounds, such as cleaning/disinfecting agents, environmental tobacco smoke, formaldehyde, and volatile organic compounds (VOCs), have been shown to be highly irritating to such areas as the eyes, skin, and upper airways. Common sources of many VOCs include solvents, multi-part forms (which use carbon copies), photocopying machines, caulking compounds, floor adhesives, and floor finishes. Certain irritants associated with accelerators, stabilizers, antioxidants, and other chemical additives found in rubber and plastic products have been shown to cause contact dermatitis. Finally, aldehydes, including acrolein and formaldehyde, VOCs, and polynuclear aromatic hydrocarbons, are among the irritating and odorous compounds produced during laser surgery.[3]

Direct toxins, mutagens, and teratogens. The hospital environment is replete with materials that, given sufficient exposure periods at high enough concentrations, will elicit direct toxic effects on people. For example, at elevated concentrations, many anesthetic agents have demonstrated toxic effects on organ systems to certain categories of HCWs. A variety of effects have been examined, including cancer, renal and hepatic disorders, nervous system effects, and reproductive effects.[4,5]

Many antineoplastic agents have been reported to be carcinogenic, mutagenic, and teratogenic.[4] At one time, it was common for HCWs to prepare the agents in areas without hoods. However, since mutagenic agents have been found in the urine of workers using horizontal laminar flow hoods,[5] this has changed. Furthermore, selection and use of appropriate gloves is essential. Nitrile has been shown to be far less permeable to many chemotoxic drugs than either the latex or polyvinyl chloride gloves used previously.

Implementation of specific work practice guidelines will minimize the potential for exposure. These recommendations include that antineoplastic agents be mixed in environmentally controlled settings by pharmacists. Furthermore, strict protocols must be developed and adhered to for the administration and disposal of antineoplastic drugs, including cleaning up spills.

Ribavirin is a synthetic nucleotide that is administered to the patient as an aerosolized medication for the short-term treatment of severe respiratory synctal virus (RSV) infections. Although ribavirin has not been linked to fetal abnormalities in humans, the teratogenic potential demonstrated in several animal species requires that avoidance of ribavirin prior to pregnancy, during pregnancy and during lactation be strongly recommended.[6] Specific engineering controls and personal protective equipment have been shown to be effective in controlling exposure.[7] (This is discussed in more detail in the "Aerosolized medications" section, page 37.)

Ethylene oxide (EtO) is used in many hospitals to sterilize medical devices and equipment. It is very effective in the sterilization of heat and moisture-sensitive items, which cannot be sterilized by steam. At high doses, however, EtO has induced dominant lethal mutations and has caused embryotoxicity in rodents. EtO in air at concentrations of approximately 200 parts per million can irritate the eyes, nose, and throat of most people. Higher concentrations can cause coughing, lung irritation, breathing difficulties, and chest pain.

Infectious aerosols. Infectious aerosols can also affect the occupants of health care facilities. Some classic examples of infection-related illnesses are legionella, (see Chapter 2), measles, chicken pox, and tuberculosis.

Other infectious diseases can be easily transmitted by ventilation systems. Cases of nosocomial infection with *aspergillus* have been reported in immunocompromised patients.[8] Nonimmune or immunocompromised individuals are easily infected with measles or chicken pox. Nosocomial acquisition of tuberculosis usually requires a host with some predisposing factors as well as prolonged or intimate exposure to infectious droplet nuclei.

Routes of Exposure

General exposure types. Although we are primarily concerned with the effects of pollutants on health, understanding the types of exposure that take place is necessary to control, reduce, and eliminate adverse impacts. There are three general exposure types of concern:

1. Building exposures (BE) are conveyed by general building systems, such as the heating, ventilating, and air-conditioning (HVAC) system, or by structural materials. Examples are entrainment of a *Legionella*-laden cooling tower mist by the outside air intake and sensitization by misapplied exterior waterproofing materials containing isocyanate.

2. Area exposures (AE) can be localized over a floor, a wing, or an area generally greater than a patient room. Examples are VOCs and/or formaldehyde from newly renovated areas, cleaning compounds that are not properly diluted, impact from nearby construction, and certain infectious diseases. Macroscale transport mechanisms (for example, HVAC recirculation, convective currents, pressure induced flows) are often involved in these cases.

3. Personal exposures (PEs) are generally of greatest significance in hospital environments and are often the least understood or characterized. People often find themselves in close proximity to air contaminant sources. Examples include administration of aerosolized medications, specific applications of cleaning compounds, donning high-allergen content latex gloves, and being near high volume photocopiers or printers. The primary characteristic of this category is that the sources are often within one meter of the person.

PEs are often the most significant exposures as well as the most difficult to characterize and control. Rodes et al provide a summary from the literature that shows that typical values of PEs would range from approximately 3.0 to 10.0 times those of AEs.[9] This indicates that the influence of personal activity sources or exposure to indoor contaminants can be significant in occupational settings, especially in those situations in which workers are in close proximity to a relatively strong source for extended periods of time. This is supported by the work of Swanson et al at the Mayo Clinic, who looked at latex allergen exposure in several settings.[10] They found that in areas of extensive glove use, the levels ranged from 13 to 208 milligrams per cubic meter (mg/m^3). They also found that within a surgical suite, there was a defined gradient between samples characterizing the room (AE) and samples taken from the operating table or from the anesthesiologist (PE). The room samples (AE) ranged from 88 to 151 mg/m^3, approximately seven to nine times lower than the PEs (549–974 mg/m^3).

Although desirable in theory, monitoring PEs in practice is often difficult or even impossible to accomplish while achieving the level of sensitivity required to make meaningful exposure determinations. Oftentimes, the required monitoring equipment is too cumbersome, heavy, or loud to be used routinely in and around patients. Sometimes, the equipment necessary to make exposure determinations has not been miniaturized or made portable enough to be useful. It is imperative that in assessing exposure, industrial hygienists have some means of quantifying the amount of uncertainty introduced by using area measurements to approximate PEs.

General exposure model. Developing a model to evaluate exposure to environmental contaminants is essential in understanding risk. The risk model presented in Figure 3–1, below, can be utilized in this assessment, allowing a more formal review of means of control in reducing risk.

The definition of terms is important in understanding this model. The *source* is defined as the presence of a material that is potentially harmful to occupants. Inherent in this is a release mechanism, such as nebulizer output in cubic centimeters per minute, that will allow calculation of a source term.

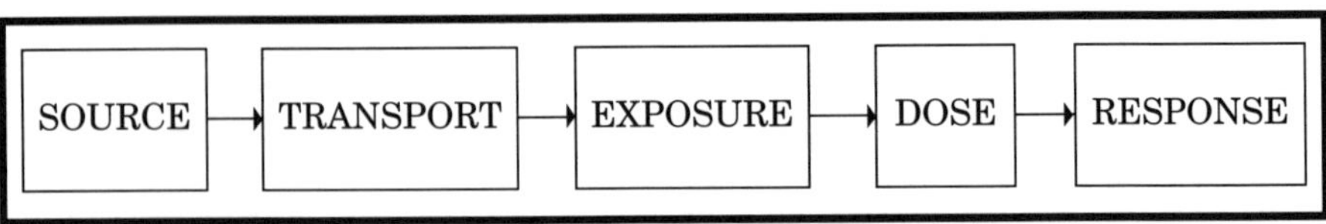

Figure 3–1: A general risk model usually follows this progression.

Transport represents the fact that a pollutant must be transported through a vector such as air, liquid, or surface contact with dust and refers to a collection of terms involving various effects, including dilution through ventilation, deposition on surfaces, transformation to other compounds, and capture. The interaction of these various mechanisms ultimately define the concentration of a pollutant at a given time.

Exposure is defined as the event of a person coming in contact with a pollutant, in contrast to environmental measurement, which is simply a measure of the pollutants in an environment. Unfortunately, an environmental measurement is often mistakenly used as a surrogate for exposure when, in fact, the simple definition given here requires the joint presence of two factors: a person and a given concentration of pollutants.

The classic toxicological truism that the "the dose differentiates a poison from a remedy" is often poorly considered in many building evaluations. If we consider transport as the presentation of a pollutant to the boundary of a receptor (person), then *dose* represents the fraction of a pollution concentration that actually passes through the boundary and is retained. This differs from exposure in that exposure occurs if the pollutant merely comes in contact with the boundary of the receptor. Examples of boundaries as used in this model include the lungs, skin, mucous membranes, and gastrointestinal tract.

The *response* is the impact of the dosage of pollutant received on the receptor's performance. As such it will be modified by an individual's pre-existing medical conditions, genetic makeup, metabolism, stress levels, prior sensitization, and exposures to various other materials occurring concurrently or sequentially.

Emerging Hazards

Latex. It is widely recognized that exposure to latex allergen can cause local reactions such as urticaria upon contact. Latex allergen can trigger type I immune mechanisms, causing allergic rhinitis, conjunctivitis, wheezing, and bronchospasm upon inhalation. With severely sensitized people, exposure to latex allergen can cause anaphylactic or systemic effects, such as tachycardia, hypotension, and bronchial constrictions.

The estimates of latex allergy risk for HCWs range from approximately 3% to 10%,[11,12] whereas that for the general population (nonHCW) are generally less than 1%. A concern remains that there was a dramatic increase in the number of HCWs who acquired sensitivity to latex proteins in the early 1990s when lower quality gloves were placed on the market to meet the increased demand required by implementation of universal precautions in 1988.

Several investigators have published detailed evaluations of latex aeroallergens in medical centers.[12,13] In these studies, high levels of latex allergen were consistently found in areas where powdered latex gloves were used in high numbers (for example, operating rooms, urology areas, orthodontics clinics, and blood bank clinics) as opposed to areas where glove use was minimal (concentrations <2 g/m^3). Personal samples were generally eight to ten times higher than area samples collected concurrently.

To control exposures effectively, hospitals should substitute powder-free, low-allergen latex gloves or nonlatex gloves for the high-allergen content gloves. Detailed cleaning of high glove use areas is also required after the glove conversion to remove reservoirs that may remain (for example, above ceilings, in and around ductwork, and on equipment surfaces) to ensure that extremely sensitive individuals can remain at work. In addition, it is important to develop latex guidelines in conjunction with the occupational health department to provide effective management of sensitized employees.[14]

Surgical smoke. Lasers are now used in many surgical applications. They can reduce trauma to surrounding tissue and promote healing by cauterizing small blood vessels. With the increased practice of laparoscopic procedures, lasers are being used more frequently for microsurgical operations, various dermatological procedures, and more recently in some dental procedures.

Lasers work by transferring electromagnetic energy into tissue; this releases a heated plume containing particles, gases, and tissue debris from the point of incision. The heating and burning of flesh releases a wide variety of acrid smelling and irritating gases and particles. Aldehydes, including acrolein and formaldehyde, VOCs, and polynuclear aromatic hydrocarbons, are among some of the odorous and irritating compounds released into the breathing zone of the surgical team. Mucous membrane irritation has also been reported from the high concentrations (that is, > 8 mg/m^3) of respirable particles that are released as well.

The debate over whether organic debris in the laser plume is harmful to HCWs has gone on for more than 10 years with no definitive resolution. Many investigators have assessed the potential for aerosolization of malignant cells, bacteria, and viruses. Bellina et al looked at the potential hazard to operating room personnel when irradiating tumors with a carbon dioxide (CO_2) laser.[15] Metabolic and cytologic examination of cellular debris detected no cytoplasmic or nuclear activity. Nezhat et al characterized the smoke plume extracted from the abdominal cavity of patients undergoing laser endoscopic treatment for endometriosis and adhesions[16] and concluded that viable cancer cells could not exist in the plume.

In contrast, Walker et al inoculated strips of human skin with viable bacilli[17] and, using both pulse and continuous CO_2 lasers, demonstrated dissemination of viable particles. This has implications for treatment of infectious lesions, as well as for dental and dermatological procedures, which are usually done in an outpatient clinic that does not have the high ventilation rates of surgical operating rooms.

Baggish, Garden et al, and Sawchuck et al provide evidence that particles, mutagens, and viral materials do exist in laser plumes.[18,19,20] Therefore, there is sufficient evidence to consider laser and electrosurgery procedures as potential risks for patients and medical personnel. It is important that appropriate smoke extraction systems be used to control the plume at the source. Furthermore, surgical teams should consider upgrading their personal protective equipment to include full face shields and improved respiratory protection (that is, a 95% efficient mask/respirator).

Anesthetic gases. Exposures to anesthetic gases is a significant concern at all medical institutions. The six anesthesia agents presented in Table 3–1, page 36, represent the major agents used in the United States

Agent	Limit	Reference
Nitrous Oxide	50 PPM	ACGIH 8 hour TLV-TWA
	25 PPM	NIOSH 8 hour REL-TWA
Halothane	50 PPM	ACGIH 8 hour TLV-TWA
	2 PPM	NIOSH 60 minute REL-C
Enflurane	75 PPM	ACGIH 8 hour TLV-TWA
	2 PPM	NIOSH 60 minute REL-C
Isoflurane	No established guideline	
Methoxyflurane	2 PPM	NIOSH 15 minute REL-C
Suprane	No established guideline	

Key:

TWA: time weighted average   TLV: threshold limit value

REL: recommended exposure limit   C: ceiling value (never to be exceeded)

ACGIH: American Conference of Governmental Industrial Hygienists

NIOSH: National Institute of Occupational Safety and Health

Table 3-1: *Major Anesthetic Agents Used in the United States and Relevant Exposure Limits*

today.

Several studies indicate that pregnant women in their second and third trimesters who are exposed to nitrous oxide at levels exceeding guideline values are at increased risk for spontaneous abortions, stillbirths, and congenital abnormalities in their children. Several epidemiological studies indicate that exposure to nitrous oxide might also be responsible for increases in hepatic and renal disease,[4,21,22] including psychomotor difficulties and hematopoietic, central nervous, hepatic, and renal system diseases and dysfunction.[23]

HCWs are typically exposed to anesthetic agents as a waste gas stream or as a fugitive gas. A *waste anesthetic gas* (WAG), for example, can be exhaled by an anesthetized patient either during or after a procedure, or it can leak from the exhaust side of the anesthesia administration apparatus. A *fugitive anesthetic gas* is gas that has escaped from any system that is functionally upstream of the patient (for example, the vaporizer/ventilator, the high pressure nitrous oxide supply system outlets).

WAGs can be effectively controlled by the proper use of a vacuum scavenging system, typically situated in the anesthesiologist's station.[23,24] WAGs that are emitted in surgical recovery units are potential sources of exposure to HCWs, especially near the breathing zone of patients immediately after their arrival in the recovery area. In this situation, general room ventilation is best for controlling exposures. Therefore, it is imperative that there be adequate amounts of outside air supplied to reduce WAG concentrations effectively, as well as to ensure that there is no "short-circuiting" or dead space in the room.

It is best to control fugitive anesthetic gases by eliminating them at the source, by tightening fittings, reducing leakage around seals, and improving work practices and by performing routine evaluations to determine the presence of these leaks. The high ventilation rates found in operating rooms (that is, 15 to 20 air changes per hour) are usually sufficient to control low-level leaks to below threshold limits. Exposure concerns arise in the area directly adjacent to the leak or if ventilation system deficiencies reduce the effective ventilation rate.

Exposure assessment involves a combination of real-time measurements, area measurements, and integrated PE monitoring. Most real-time measurements for nitrous oxide and the halogenated anesthetic gases involve long pathlength infrared spectrophotometry or photoacoustic infrared spectrophotometry. These measurements can be useful in determining fugitive emissions from various sources as well as determining exposure profiles during procedures.

Active (pumped) and passive dosimetry systems can both measure nitrous oxide and halogenated anesthetic gases. These can be used to determine time-weighted averages in rooms. They are especially useful in assessing PEs of HCWs to these agents over extended time periods (2 to 12 hours).

Aerosolized medications. Many medications are now being aerosolized and administered via inhalation because of increased efficacy in treating various diseases. Ribavirin, pentamidine, amphotericin, colistin, DNAse, gentomycin, and tobramycin, among others, are given routinely in many hospitals across the United States. Although much information is available regarding efficacy of this type of medication as treatment, there is generally little information regarding impacts of exposure on HCWs. Control of aerosolized medications is worthy of additional discussion.

Ribavirin is an antiviral agent used for the treatment of RSV infection. The drug has been shown to be teratogenic in some animal species[25,26,27]; therefore, detailed policies regarding administration have been developed to minimize HCW exposure. No occupational exposure standard has been recommended by OSHA, NIOSH, or the American Conference of Governmental Industrial Hygienists (ACGIH). The California Department of Health Services has recommended a limit of 2.7 lg/m^3 as an eight-hour time-weighted average based on taking the No Observed Effect Level in the most sensitive animal species and applying a safety factor of 1,000.[28]

Minimization of HCW exposure requires that a combination of administrative and engineering controls be employed. Ideally, a fully enclosed scavenging system will minimize release into the general room environment. All rooms in which ribavirin is administered should conform to American Institute of Architects guidelines for total air exchange rates and be negative with respect to adjoining spaces during and following administration. People entering the room should wear appropriate personal protective equipment, including high efficiency particulate arrestance (HEPA) filtered respirators, nonsterile gloves, a disposable jumpsuit, hair cover and foot covers, and goggles.

Most importantly, health care facilities must provide appropriate training to HCWs to educate them about the potential risks of ribavirin and the appropriate methods to control and minimize exposures.

Other aerosolized medications do not have the same health effects data that ribavirin does. Very little is known about the mutagenic or teratogenic effects of certain medications. However, anecdotal reports indicate that some HCWs experience symptoms of scratchy throat, burning eyes, rashes, and irritation and exacerbation of certain allergies under prolonged exposure to these aerosols.

As with ribavirin, appropriate engineering controls can be very effective in limiting exposure to other aerosolized medications. Until a specific medication can be conclusively ruled out as causing health effects, we advocate taking a cautious approach. We recommend that nebulizers be utilized that have absolute filters on their exhaust leg. We also recommend that HCWs wear N–95 respirators or better, as well as gloves, eye protection, and precaution gowns when administering aerosolized medications.

In contrast with SBS and BRI, which can be controlled through maintenance or reengineering, occupational exposures require ongoing vigilance and education to keep them to a minimum.

References

1. Kern DG, Frumkin H: Asthma in respiratory therapists. *An Intern Med* 110:767–73, 1989.

2. Chistiani DC, Kern DG: Increased prevalence of asthma in respiratory therapists. *Am Rev Respir Dis* 143:A441, 1991.

3. Ott D: Smoke production and smoke reduction in endoscopic surgery: preliminary report. *End Surg* 1:230–32, 1992.

4. Selevan S, Lindholm ML, et al: A study of occupational exposure to antineoplastic drugs and fetal loss in nurses. *N Engl J Med* 313:1173–78, 1985.

5. Anderson RW, Puckett WH, Dana WJ, et al: Risk of handling injectable antineoplastic agents. *Am Soc Hosp Pharm* 39:1881–87, 1982.

6. Waskin H: Toxicology of antimicrobial aerosols: A review of aerosolized ribavirin and pentamidine. *Respiratory Care* 36:1026–36, 1991.

7. Health Hazard Evaluation Report HETA 91–104–2229: Florida Hospital: Orlando, Florida. Cincinnati: NIOSH, 1992.

8. Rhame ES: Prevention of nosocomial *aspergillus*. *J Hospital Infection* 18(A):466–72, 1991.

9.   Rodes CE, Kamens RM, Wiener RW: The significance and characteristics of the personal activity cloud on exposure assessment measurements for indoor contaminants. *Indoor Air* 2:123–45, 1991.

10.   Swanson MC, Bubak ME, Hunt LW, et al: Quantification of occupational latex aeroallergens in a medical center. *J Allergy Clin Immunol* 94:455–51, 1994.

11.   Medical Alert Bulletin U.S. Food and Drug Administration. Mar 29, 1991.

12.   Grzybowski M, Ownby DR, Peyser PA, et al: Prevalence of anti-latex IgE antibodies among registered nurses. *J Allergy Clin Immunol* 98:535–44, Sep 1996.

13.   Heilman DK, Jones RT, Swanson MC, Yuninger JW: A prospective, controlled study showing that rubber gloves are the major contributor to latex aeroallergen levels in operating room. *J Allergy Clin Immunol* 98:325–30, 1996.

14.   Sussman G, Gold M: *Guidelines for Management of Latex Allergies and Safe Latex Use in Health Care Facilities.* Arlington Heights, IL: American College of Allergy, Asthma & Immunology, 1996.

15.   Bellina JH, Stjernholm RL, Kurpel JE: Analysis of plume emissions after papovavirus irradiation with the carbon dioxide laser. *The Journal of Reproductive Medicine* 27:268–70, 1982.

16.   Nezhat C, Winer WK, Nezhat F, et al: Smoke from laser surgery: Is there a health hazard? *Lasers in Surgery and Medicine* 7:376–82, 1987.

17.   Walker NPJ, Matthews J, and Newsom SWB: Possible hazards from irradiation with the carbon dioxide laser. *Lasers in Surgery and Medicine* 6:84–86, 1986.

18.   Baggish MS, Poiesz BJ, Joret D, et al: Presence of human immunodeficiency virus DNA in laser smoke. *Lasers in Surgery and Medicine* 11:197–203, 1991.

19.   Garden JM, O'Bannion K, Sheinitz LS, et al: Papillomavirus in the vapor of carbon dioxide laser treated verrucase. *JAMA* 259:1199–202, 1988.

20. Sawchuck WS, Weber JP, Lowry DR, et al: Infectious papillomavirus in the vapour of warts treated with carbon dioxide laser or electrocoagulation: Detection and protection. *J Am Acad Derm* 21:41–49, 1989.

21. Rogers B: A Review of the toxic effects of waste anesthetic gases. *Journal of the American Association of Occupational Health Nurses* 34(12):574–79, Dec 1986.

22. Cohen EN, Bellville JW, Brown BW: Anesthesia, pregnancy and miscarriage: A study of operating room nurses and anesthetists. *Anesthesiology* 35:343–47, 1991.

23. Gilly H, Lex C, Steinbereithner KL: Anesthetic gas contamination in the operating room an unsolved problem? *Anaesthesist* 40(11):629–37, Nov 1991.

24. Burkhart JE, Stobbe TJ: Real-time measurement and control of waste anesthetic gases during veterinary surgeries. *Amer Ind Hyg Assoc J* 51:640–45, 1990.

25. Kochar DM, Penner JD, Knudsen TB: Embryonic, teratogenic and metabolic effects of ribavirin in mice. *Toxic Appl Pharmacol* 52:100–112, 1980.

26. Fernandez H, Banks, G, Smith R: Ribavirin: A clinical overview. *Eur J Epidemiol* 2:1–14, 1986.

27. Kilham L, Ferm VH: Congenital anomalies induced in hamster embryos with ribavirin. *Science* 195:413–14, 1997.

28. California Department of Health Services. *Health Care Worker Exposure to Ribavirin Aerosol: Field Investigation FI–86–009.* Berkeley, CA: California Department of Health Services, Occupational Health Surveillance and Evaluation Program, 1988.

Chapter 4: Patient Impact

Irene Orme, RN, CIC

Ms Orme is President, Orme Healthcare Consultants, West Hills, California. Ms Orme has been active in nursing for more than 35 years. She has more than 20 years of experience in dealing with regulatory issues in the field of infection control and epidemiology. She has twice served as a president of the Greater Los Angeles Chapter of the Association of Practitioners in Infection Control and Epidemiology (APIC), and has chaired the California APIC Coordinating Council.

"Epidemic fever is caused by the air. Because all men inhale the same wind, when the air is infected with such pollutions that are hostile to the human race, the men fall sick."
—Hippocrates
460 377 BC

The primary business of health care facilities is to provide patient care. The very nature of this business can expose both patients and employees to more infectious, allergenic and other pollutants than people are exposed to in non-health care settings. The impact of airborne pollutants on patients differs from the impact on employees in several areas:

- Patients are at greater risk because their underlying illness or condition renders them susceptible to diseases that may not affect a normally well person.

- Patients are at greater risk to pollutants because normal host defenses have been breached by procedures and treatment that leave them more vulnerable to attack.

- Patients have increasingly shorter lengths of stay, thus reducing exposure time to pollutants.

- Patients are not usually exposed to cleaning, disinfecting, preserving, and other substances used to maintain the physical plant and equipment.

Direct Effect of Airborne Contaminants

Infectious contaminants transmitted by the airborne route are most commonly fungal, bacterial, and viral. Although infections acquired by the airborne route make up a relatively small portion of nosocomial, or hospital acquired infections, the outcome to the patient of these infections can be grave. As described in Chapter 2, the two most serious and prevalent diseases resulting from airborne transmission are legionellosis (Legionnaire's disease) and aspergillosis.

Airborne Bacterial Diseases

The severity of disease caused by aerosols of bacteria depends on several factors:

■ The virulence of the particular organism;

■ The size of the innoculum, or the level of bacteria present;

■ The underlying disease or health status of the patient; and

■ The ventilation and movement of air in the patient's environment, or the exposure of the host to the organism.

Given these factors, the most serious of bacterial infections is legionellosis. There are more than 30 species of *Legionella,* the most harmful being *Legionella pneumophila.* Typically, *Legionella* species are not identified on routine cultures, giving rise to the suspicion that the disease is underreported.

The most severely affected patient population is those who are highly immunosuppressed. Outbreaks are often clustered in a specific group of patients, such as bone marrow transplant recipients, solid organ transplant recipients, patients with immunodeficiency diseases such as AIDS, the elderly and neonates. Antibiotic therapy is effective if the disease is recognized and treated early.

Although other bacteria may become airborne—especially *Staphylococcus, Pneumonococcus,* and *Streptococcus*—airborne nosocomial transmission has not been reliably documented. However, this should not be ruled out entirely.

Airborne Viral Diseases

Outbreaks of viral disease transmission in health care settings, in contrast, have been documented at length. Measles and varicella (chicken pox) are known to be transmitted by the airborne route. Both of these diseases are usually introduced into the health care setting in an uncontrolled manner and it is possible if not probable for an ambulatory contagious patient to expose others in several different departments within the facility.

Chicken pox is probably the most contagious of childhood diseases, but the public perception is that it is a mild disease of little consequence. Pregnant mothers have brought infected children with them on prenatal appointments for lack of a baby-sitter. Infected persons have been sent to radiology and laboratory departments for diagnostic procedures. Secondary transmission of varicella has been documented in patients in rooms down the hall from the infected person,[1,2,3] or in persons passing the open door of an infected patient in a room without negative pressure ventilation.

Measles is the most contagious prior to the skin rash phase. Secondary measles transmission has been documented in physician's offices up to 90 minutes after a coughing, infected child has left the room.[4,5,6]

Contrary to popular opinion, the impact of both of these diseases on patients can be life threatening. Mortality is the highest in pediatric populations, especially if the child is malnourished or immunocompromised. Both diseases are of increased severity in normal adults and life threatening to immunocompromised adults. Fortunately, most adults have immunity to measles and varicella.

Fungal Infections

Aspergillus, a fungi plant, plays an important part in survival on the planet. Without *Aspergillus,* decomposition would not occur. Fallen trees in the forest would not rot and fallen leaves and other debris would not decompose; but the problem that it creates in a health care setting is well known (see Chapter 2).

Aspergillosis is acquired by inhalation of airborne dust particles that carry the spores. Pneumonia develops and the fungus disseminates through the blood stream to other organs. Mortality rates have been reported as high as:

- 95% in bone marrow transplant patients;[7]

- 13–80% in leukemia patients;[8,9] and

- 8–30% in kidney transplant patients.[10,11]

Construction management is crucial to fungal contamination prevention and is discussed in detail in Chapter 7, page 88; but we discuss why this is necessary here. One large transplant referral HMO found that the mortality rate in liver patients declined from a probable 13% to no documented cases the next year when the patients were sent from a health care facility undergoing renovation to a facility where no construction and renovation was taking place. Another tertiary care center noted a reduction from a 9% rate in aspergillosis to no documented cases the following year after the development and enforcement of an extensive dust containment policy during construction and renovation, which included inspections by the infection control team and serious fines for noncompliant contractors.

However, despite the use of anti-fungal drugs, the outcome of transplant patients with aspergillosis continues to be grim, due to the underlying disease or condition and the therapy given to induce immunosuppression as a means to prevent rejection of the transplanted organ.

Patient Impact from Allergenic Contaminants

Allergens are normally harmless substances that can cause a reaction in sensitive people. Health care facilities harbor many allergens, but documentation of nosocomial outbreaks is rare. Environmentally related allergic reactions depend upon these three interactive elements:

1. Exposure to the allergen;

2. Individual susceptibility; and

3. Length of time exposed.

Symptoms of allergic reactions such as runny nose, sneezing, coughing, and itching and watery eyes may be attributed to other diseases or conditions. It is suspected that sensitive, exposed patients are frequently discharged before an accurate diagnosis is made.

Microorganisms that cause allergic reactions. The most common microorganisms that cause allergenic reactions are environmental molds and bacteria. Molds can be found in damp spots such as shower stalls, water-stained ceiling tiles, mattresses, and foam rubber pillows. The spores become important only if they become airborne. Agitation of pillows and mattresses (and shower mist) can send off a bout of sneezing and coughing in some individuals.

The National Institute of Occupational Safety and Health reports a high level of both molds and bacteria, especially *bacillus* spp. in air samples taken from health care facilities. Fibrous insulation materials taken from air ducts where allergic complaints have been identified have also been documented.

Endotoxins, a cell wall product of the metabolism of gram negative bacteria, may cause a hypersensitivity reaction in some individuals and can be an issue in the health care environment. Allergic reactions to animals is common but is not a problem in the health care setting. However, the waste products of cockroaches and dust mites, and dead dust mites in the air, do cause an allergic reaction in the general population and severe reactions in immunosuppressed patients. The waste products, which are proteins, are usually part of the mixture of potentially allergenic materials that make up dust. House dust-mite allergy is the most common year round allergy in the world.

Nosocomial infestation of patients by avian mites has also been documented. In a San Diego hospital, mites from nearby roosting pigeons entered the building via air-conditioning ducts.[12] Mites are attracted to heat, and when avian hosts are scarce, they seek out and attack humans.

Air Quality Related to Patient Comfort and Convenience

Thermal concerns. Patient complaints related to the temperature of their environment are among the most common complaints voiced in a hospital. The patient's condition decreases his or her ability to adequately maintain homeostasis, and his or her thermoregulatory system malfunctions. Diagnostic and treatment procedures often alter body temperature or the patient's ability to regulate it. As the body attempts to regulate the internal temperature, fevers followed by chills and sweats can occur. The ASHRAE recommendations of 73°F–79°F may or may not be comfortable for the patient undergoing these homeostatic altering treatments. Plant service engineers need to be cognizant of the importance of eliminating drafts and adjusting ambient room temperatures as part of the overall effort to improve patient comfort.

Relative humidity (RH). The RH comfort range recommended by ASHRAE is 30% to 60% RH. The validity and importance of this is shown in Chapter 6, Figure 6–3, page 70. Although warmer air holds more vapor, heat that is added to indoor air in the colder months has a drying effect and lowers the indoor RH unless an indoor moisture source is present. The American Academy of Otolaryngology recommends maintaining a bedroom RH of 45% to avoid the risk of respiratory problems and sore throats. Some believe that dry air, or low RH, can possibly increase the risk of respiratory problems, but these findings have not been supported by studies and remain controversial. Dry rooms can cause a patient to feel colder (especially after a bath) because the human body cools through evaporation when exposed to dry air.

Humidity higher than 60%, however, has been shown to increase risk of respiratory ailments because it promotes the growth of airborne fungi, bacteria, molds and other allergens and may even favor the survival of some viruses.

Fumes and noxious odors. Fumes and noxious odors affecting patients are more likely to be sentinel events rather than an ongoing, constant factor in the environment. Fumes are formed when material from a volatized solid condenses in cool air. Patient complaints can range from nuisance odor complaints by allergic or hypersensitive patients to complaints of fumes and odors that could result in illness. Complaints of odors emitting from other patients, those of fragrances, perfumes, and even of room deodorizers are not usually a health threat, but resolution of the problem is important to the patient's well-being. Fumes and vapors such as ammonia inhalation, which cause irritation to the lining of nose, throat, and lungs are a health hazard.

Some fumes, while not necessarily a health hazard for short-term exposure, are so noxious that evacuation of the area may be required. For example, a contractor using tar to repair an earthquake-damaged hospital roof in the Los Angeles area failed to cover adjacent air intake ducts, resulting in complaints from patients and employees on the floor directly below the repair site. Palliative measures such as increased ventilation and using large industrial fans after covering the air intake ducts did not alleviate the problem. The patients had to be relocated for their comfort.

Noxious odors in an ambulatory care center in Santa Clarita, California, forced occupants to evacuate the building and the center to close until the source of the odors was found and eliminated. The problem originated in the return air plenum that was common with other tenants in the building. Because of rodent complaints, the adjacent space's tenants had spread rat poison in the plenum. The odor of decomposing rodents in the plenum permeated the medical facility. It was necessary to close and evacuate the facility while the HVAC system, return air plenum, and some wall cavities were cleaned and decontaminated.

Prompt investigation of all complaints of fumes and odors and early intervention may require nothing more than an explanation from an authoritative person to reassure the patient that there is no health hazard.

Airborne Problems Following Natural Disasters

Earthquakes. Air pollutants in a medical facility following an earthquake are usually related to the structural damage sustained. Broken sewer and water pipes spill their contents into the environment creating water damage and a potential risk of disease. Water stained ceiling tiles, wall board, upholstered furniture, and carpets become a reservoir for molds and other allergens that can become airborne (See Chapter 2, Table 2–4, page 22).

Hurricanes and tornadoes. Cyclones, hurricanes, tornadoes and other violent and destructive windstorms create facility damage similar to earthquakes, often accompanied by torrential rains. A hospital in Charleston, South Carolina, reported that Hurricane Hugo blew open or fractured windows and the sheet rock walls moved so much that they generated a thick dust.[13] Disease-causing airborne pollutants not normally found indoors, such as coccidioides or histoplasma, can enter the facility. Elevated concentrations of airborne asbestos can occur if asbestos-containing materials are disturbed. Personal protective equipment including dust mist respirators should be available for personnel working in these areas. If airborne asbestos is present, the area should not be entered until it has been cleaned by personnel specifically trained and equipped with appropriate respirators.

Floods and other disasters causing water damage. Indoor air quality may appear to be the least of problems in a hospital following a flood, but standing water is a breeding ground for many micro-organisms that can become airborne and thus be inhaled. If the flood water is contaminated with raw sewage, there is a risk of infectious disease transmission. Water damaged materials that are porous and take a long time to dry should be removed and replaced with new materials.

The cleanup process after a flood or other disasters is often extensive and disinfectants and sanitizers containing toxic substances are used. Cleanup crews must carefully read and follow label instructions and good ventilation must be provided. Fans and other means of providing good ventilation may need to be brought in.

It is imperative for patients to have good indoor air quality for the duration of their stays in health care facilities. From noxious odors, which are little more than a nuisance, to *Aspergillus,* which can kill, air quality affects the perception of care a patient receives.

References

1. Leclair JM, Zaia JA, Levin MJ: Airborne transmission of chickenpox in a hospital. *N Engl J Med* 302: 450, 1980.

2. Josephson A, Gomberrt ME: Airborne transmission of nosocomial varicella from localized zoster. *J Infect Dis* 158: 238–41, 1988.

3. Gustafson TI, Lavely GB: An outbreak of airborne nosocomial varicella. *Pediatrics* 70:550–56, 1982.

4. Remington PL, Hall W, Davis, IH, et al: Airborne transmission of measels in a physician's office. *JAMA;* 253:1574–77, 1985.

5. Bloch AB, Orenstein W, Ewing WM, et al: Measles outbreak in a pediatric practice: Airborne transmission in an office setting. *Pediatrics* 75:767–83, 1985.

6. Atkinson WL, Maskowitz LE, Adams NC, et al: Transmission of measels in medical settings: United States 1985 1989. *Am J Med* 91:320–24S, 1991.

7. Pannuti CS, Gingrich RD, Pfaller M, Wenzel R: Nosocomial pneumonia in adult patients having bone marrow transplant. *Cancer* 69(11):2653–62, 1992.

8. Denning DW, Stevens DA: Antifungal and surgical treatment of invasive Aspergillosis: Review of 2121 published cases. *Rev Infect Dis* 12:1147–201, 1990.

9. Weinberger M, Elattaor I, Marshall D, et al: Patterns of infection in patients with aplastic anemia and the emergence of *aspergillus* as a major cause of death. *Medicine* 71:24–43, 1992.

10. Torre-Cisneros J, Lopez O et al: Characteristics of organ transplant recipients with CNS *aspergillosis. J Neural Neurosurg Psychiatry* 56: 88–93, 1993.

11. Klimowski LL, Rotstain C, Cummings KM: Incidence of nosocomial aspergillosis in patients with leukemia over a 20 year period. *Infect Control Hosp Epidemiol* 10(7): 299–305, 1989.

12. Vargo J, Mizrahi M, Ginsberg M: Nosocomial infestation of patients

by pigeon mites. *State of California Department of Health Services, California Morbidity,* 1982.

13. Russell CK: Managers ask and answer. *J Emergency Nursing* 16(4):292–95, 1990.

Chapter 5: Hospital and Employee Impact

Irene Orme, RN, CIC, and Charlotte G. Lee, RN, CIC

Irene Orme, RN, CIC, is President, Orme Healthcare Consultants, West Hills, California. Ms Orme has been active in nursing for more than 35 years. She has more than 20 years of experience in dealing with regulatory issues in the field of infection control and epidemiology. She has twice served as a president of the Greater Los Angeles Chapter of the Association of Practitioners in Infection Control and Epidemiology (APIC), and has chaired the California APIC Coordinating Council.

Charlotte G. Lee, RN, CIC, is Director of Infection Control and Epidemiology, Kaiser-Permanente Medical Center, Woodland Hills, California. Ms Lee has more than 25 years of experience in the nursing and infection control fields, and 14 years of experience in occupational health and safety.

In this age of value analysis teams, cost-containment programs, and downsizing, hospitals are faced with increasing costs to build, upgrade, and maintain facilities that must conform to increasingly complex standards from a multitude of regulatory agencies. To add to this financial burden, there is the constant demand to improve patient care, and to increase health care benefits.

The financial demands on a health care facility create stress that is felt by all health care employees. To perform their jobs well in this stressful environment, health care professionals must feel comfortable and be assured of a healthy environment. Health care professionals are at greater risk of airborne pollutants and infections than the population in general. Chapter 3 lists the types of contaminants found in hospitals; it is important to understand the real effect that this exposure has on the general health of the health care organization's employees. Facility managers must react to employee complaints of IAQ problems consistently and appropriately. Employees and staff generally ask themselves two questions regarding how well they perceive their complaint to have been received:

1. *Are you listening?* Regardless of the type of business we are in, or the job title we hold, our primary responsibility is communication. Frequent employee complaints of real or perceived airborne pollutants should be addressed immediately. Complaints ignored or not taken seriously become an increasingly sensitive issue that often develops into a "we versus they" scenario. If this occurs, employees may view with suspicion subsequent action taken by administration or by those perceived to be in authority.

2. *Are you going to do something about it?* Members of the IAQ team (see Chapter 6) should make an on-site visit as soon as possible, preferably accompanied by the complainant. The initial walk-through may identify the cause of the complaint and effective resolution may be made at that time. This is usually not the case, however; complaints of airborne pollutants are not often that easy to solve and require more information. Although complaints may result from other causes, such as job-related stress, studies show that symptoms can be exacerbated by indoor pollutants. Complainants are more likely to perceive that they are receiving serious attention and meaningful action is being taken if they understand the authority of the IAQ team and feel they are being included in the IAQ team's investigation. Rather than act upon statements like, "everyone is affected" or "we're all sick," the IAQ team should use questionnaires such as those in Appendix D. A hypothesis can be developed only after these data are collected and analyzed.

Occupants of the affected area should be kept informed of action plans as they are developed and put into place. Follow-up questionnaires should be filled out after remedial action has been taken to evaluate the effectiveness of the action.

Nature of Complaints

Occupants of health care facilities are subject to many unique IAQ threats, which fall into the following four categories: biological microorganisms, mold and fungal microorganisms, chemicals and pharmacological compounds, and fumes and odors.

Biological microorganisms. As with patients, measles and varicella are the two most common viral diseases transmitted to employees by the airborne route. All health care workers *should* have immunity to these diseases; but in an imperfect world, this is not always the case. Therefore, all health care facilities should have infection control policies prohibiting susceptible people from entering the rooms of patients with measles and varicella. The infection control policies should also require patients with these diseases to be placed in rooms with negative air pressure. Despite the prevalence of these policies and precautions, 4% of all cases of measles in Los Angeles County occurred among acute care hospital employees in the three-year period 1987–1989.[1] Nine of 25 cases of measles reported in May 1996 in Clark County, Washington,[2] were medical workers. Not all of the Clark County medical workers had direct patient contact; some were exposed by the airborne route. Health care workers who are susceptible to measles and varicella are at high risk of acquiring the disease because of the close personal contact required in caring for patients.

Environmental mold and fungal microorganisms. Chapter 2 provides detailed information on environmental mold and fungal microorganisms. Although nosocomial Aspergillosis is of great concern, the increase of occupational asthma is a potential sleeping giant. The annual increase in asthma incidence[3] is shown in Figure 5–1, page 56. (This graph represents year to year changes and not the number of reported cases.)

The incidence of pediatric asthma is also on the rise, and at a faster rate than adult asthma. This is of interest and concern to pediatricians and allergists, who are still evaluating the data in an attempt to determine the cause for this increase. Of more specific interest to us here, is

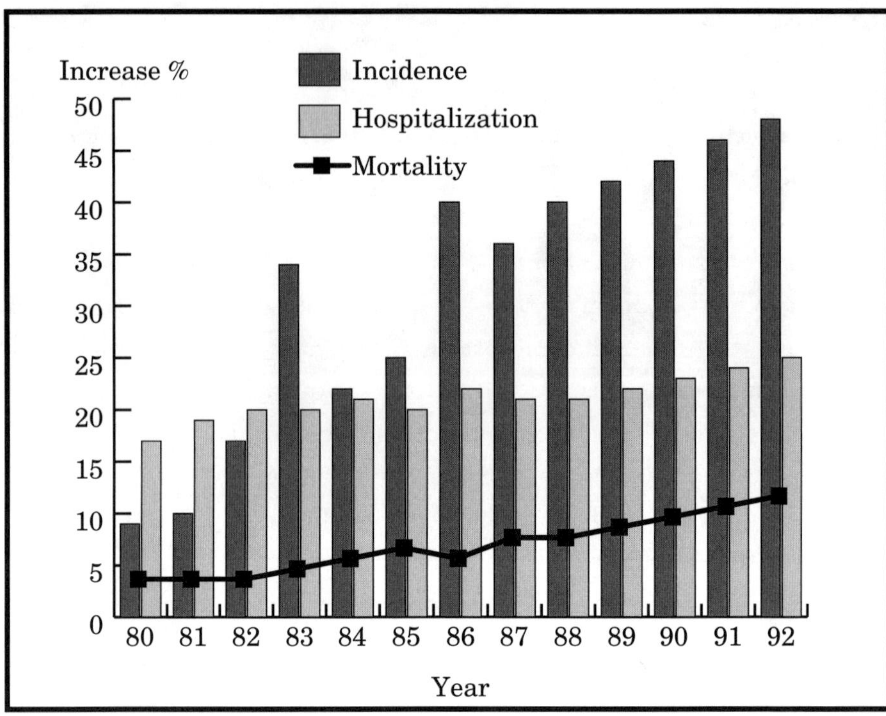

Figure 5–1: *Hospitalization and mortality for incidence of adult asthma has increased in the past 17 years.*

the increase in asthma-related hospitalization[4] and mortality[5] associated with the rise in incidence (shown graphically in Figure 5–1). As an adjunct to the human suffering aspect of asthma, the cost of treating asthma, without any consideration of the ancillary costs, has grown steadily: from $4 billion in 1986, to $6.2 billion in 1990[6], to just under 2% of all health care costs in 1996.[3]

One of the significant costs associated with IAQ-related issues, is the workers' compensation insurance claims. Figure 5–2, page 57, shows the annual increase in claims filed for the category of lung and respiratory complaints.[7] With a shoulder shrug, you might ask "How expensive could this be?" The cost of a typical claim is sobering: The national median average total claim is nearly $28,000.00.[7]

The IAQ and asthma issue has not escaped the notice of the National Institute of Occupational Health and Safety. In 1995, under the SOAP

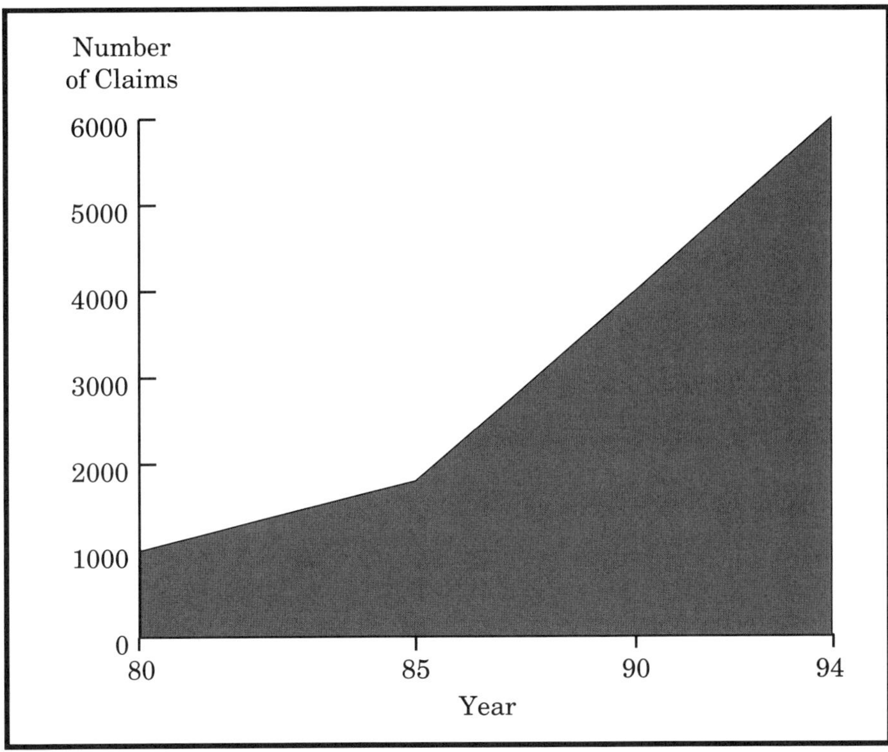

Figure 5–2: *The increase in lung and respiratory claims points to the growing problem of poor indoor air quality.*

and SENSOR programs, the institute commissioned a study on environmental asthma,[8] with an emphasis on *environmental.*

Chemicals and pharmacological compounds. The adverse effects of exposure to the vapor of glutaraldehyde (see Chapter 3) have been well documented. Glutaraldehyde is an effective germicide widely used to disinfect surgical and delicate heat-sensitive instruments. Exposure to these vapors and, to a lesser extent, the vapors of formaldehyde and formalin can generate a high level of complaints and anxiety. Complaints from employees experiencing physical distress and from those who fear long-term effects, adverse pregnancy outcome, or fetal damage can become an emotionally charged issue. Employees exposed to these vapors can come to believe that they are being poisoned.

Fumes and odors. Health care workers, as they go about the business of cleaning, disinfecting, and maintaining the physical plant and equipment, dispensing aerosolized drugs and anesthesia gases, and providing patient care are exposed to many fumes and odors.

The IAQ team can quickly investigate and resolve employee complaints of fumes and odors when sources are identified, such as those resulting from new carpeting, cleaning agents, paint refurbishing, or human bioeffluents. But some complaints appear to be ongoing with no resolution and their sources are identified only by chance, if at all.

Handling IAQ Complaints

It is necessary to prevent the escalation of IAQ complaints. Administrative, engineering, and work practice controls must be put into place, and personal protective equipment made mandatory. An example of these controls for glutaraldehyde follow:

A. Administrative Controls

- Limit employee access to glutaraldehyde by centralizing usage.

- Use only in a large well-ventilated area.

- Develop spill policy.

- Install and operate proper ventilation and exhaust equipment.

- Use tight-fitting lids on disinfecting trays.

- Use automated scope disinfecting equipment if possible.

- Use automated systems using alternative to glutaraldehyde.

B. Work Practice Controls

- Keep lids on disinfecting trays, except when actually transferring instruments into and out of trays.

- Use a pick up instrument to transfer instruments in and out of solution and create as little agitation as possible.

- Take particular care when passing solution into or out of trays to reduce aerosolization.

- Wipe up spills immediately and properly dispose of rags and towels by rinsing with water and sealing in plastic bags for disposal.

C. Personal Protective Equipment

■ Use eye protection.

■ Use glutaraldehyde impervious gloves.

■ Use protective clothing if splashing is likely.

Occasionally one encounters a casual or cavalier attitude among employees. Many health care workers, especially those in ambulatory care settings and physician offices, see their primary function as delivering patient care and assisting the physician. They may see processing and disinfecting equipment, and processing and packaging specimens to be delivered to a laboratory, as tasks to be accomplished in their spare time with little or no thought given to the toxicity of the materials being handled. They may have received little or no training; for example, without a written procedure for training new employees, some may have received only hurried instructions from the departing employee. The same standard of practice must be observed in all health care settings. Just as hospital health care workers must have documented department-specific training, so must health care workers in other settings.

Faced with this information, it clearly becomes evident that to have a definitive IAQ program in place is in the health care organization's best interest. A proactive approach to problem solution or avoidance is always the most cost-effective and least painful method of dealing with any issue, including IAQ. Although a great deal of responsibility for solving facility-related problems rests with the facilities engineering departments, infection control and employee safety departments must also take an active part in a facility's preventive maintenance program.[9]

As an example of what typically happens when a proactive program is not in place, consider this episode at a Southern California hospital in 1996. An employee perceived that his general ill health was building-related. This led to more employees complaining as unauthorized pictures of internal components of the medical center HVAC system were circulated. Occupational health, plant operations and infection control personnel formed a team. The team developed a systematic plan to address the employee concerns and determine what effect the HVAC system actually had on employee health and safety. Members contacted an outside IAQ consultant to conduct a professional assessment and assist with the investigation. While these actions were in progress, the complainant filed a union grievance and a workers' compensation claim.

The action steps that the team developed specifically targeted all areas housing employees with complaints. The IAQ consultant examined the HVAC system and took culture samples from diffusers, registers, ducts, coils and filters. Air samples were obtained from the work space of each of the now expanded list of complainants. The results of all tests done supported the conclusion that the index employee's complaints were unfounded and the IAQ was in fact above regulatory requirements. The IAQ team gave all peripherally involved employees "black and white" documentation to prove that their concerns had been adequately addressed. The complainant withdrew his compensation claim, and no additional claims were filed. These steps were correct and thorough; but in performing them on a reactive rather than proactive basis, they became much more costly to the hospital in terms of out-of-pocket consultants fees, staff time spent, involvement of the regional employee safety and industrial hygiene personnel, and the general stress for those involved.

References

1. Los Angeles County Department of Health Services: *Public Health Letter* 12(1), 1990.

2. Washington State Department of Health: *Department of Health Press Release:* 96-52, May 1996. Http://www.deh.wa.gov/publicat/96–42.html.

3. Fact sheet: National Asthma Education Program, Bethesda, MD.

4. Data Fact Sheet: National Center for Health Statistics, National Health Interview Survey, 1994.

5. Sly RM: Mortality from asthma. *Journal of Allergy and Immunology* Nov: 1988.

6. Update: Americans spend $6.2 billion on asthma care, study says. National Institute of Health, and *National Institute of Allergy and Infectious Diseases* Mar 25, 1992.

7. The State of California: *Insurance Commissioner's Report,* 1993.

8. Williams RC: Federal occupational asthma program builds diagnostic model. *Indoor Air Review* Nov 1995.

9. Hansen W: Controlling IAQ in the transmission of disease. Paper presented to Association of Practitioners in Infection Control and Epidemiology (APIC), Jun 1994.

Part II: The IAQ Program in Action

Chapter 6: Developing the IAQ Program

Wayne Hansen, PE, REA, CEM,

Mr. Hansen is Director of Engineering, NIAQ™ Division, Mintie Corporation, Los Angeles, California. He has been a registered engineer for more than 25 years, and has been active in the professional indoor air quality community for 11 years. He currently serves on the Indoor Air Quality Advisory Board for California OSHA.

Why is an IAQ program needed?

As we learn more about IAQ and its effect on people, we have an obligation to use that knowledge constructively to improve the safety, health, and comfort of health care workers and patients. At this point, we have five chapters worth of reasons why a health care organization needs an IAQ program:

1. SBS is largely a perceived issue affecting the comfort of everyone in the facility. Although not a health threat in a serious sense, it is a definite problem for the facility when dealing with employees (Chapter 1).

2. BRIs are health risks and do cause clinically diagnosed diseases. Environmental mold and fungi are often the culprits in spreading these diseases (Chapter 2).

3. The general VOCs used in a health care setting, latex gloves, and the various infections health care workers are exposed to in the course of their work can create untold problems for the hospital, staff, and employees (Chapter 3).

4. Nosocomial infections are always a great concern for patients, and health care workers are not immune to the infectious agents responsible. Odors can become a source of irritation not only to those exposed, but to those whose duty it is to remove them (Chapter 4).

5. Allergies and asthma are becoming more serious emerging issues, especially for health care workers. This can have serious financial impact in terms of medical treatment costs and workers' compensation

settlements (Chapter 5).

For the person suffering, it doesn't make any practical difference whether it is from SBS or BRI. His or her only concern is that the problem be corrected so that he or she can find relief. Any new issue affecting the workplace, such as IAQ issues, may find resistance to change, but new information dictates that these changes be made.

The increased risk of exposure to disease and environmentally related maladies of health care workers has been documented by several sources, resulting in the National Institute of Occupational Safety and Health's development of a guide specifically geared to them, as shown in Table 2, page xv, of the Introduction.

Establishing an IAQ Program: What Is Involved?

After the decision is made to create and implement a comprehensive IAQ program, the next question is, "How do we get started?" Setting up a program specific to the organization is rather straightforward, with well-defined action steps:

- Organize the IAQ program team, and designate one person to be in charge of the program. This designated person can either be a specific individual, or a specific job title such as Safety Officer or Chief Engineer. Giving one person such a title gives him or her some authority and even a sense of greater responsibility for carrying out the IAQ program. The team structure might follow that shown in Figure 6–1, page 67.

- Gather all mechanical system- and facility-related documents such as building plans, submittal data, equipment lists, air balancing reports, and material data safety sheets (MSDS). Include equipment maintenance logs as they relate to IAQ in this section. (Examples of this type of log form[1] can be found in Appendix C, page 137).

- Write a short description of the facility, including the original construction date and the dates of all additions and expansions. Include a description of the current heating, ventilating, and air conditioning (HVAC) system and its major components.

■ Write an action plan for dealing with IAQ-related complaints. (Sample interview forms can be found in Appendix D, page 143.)

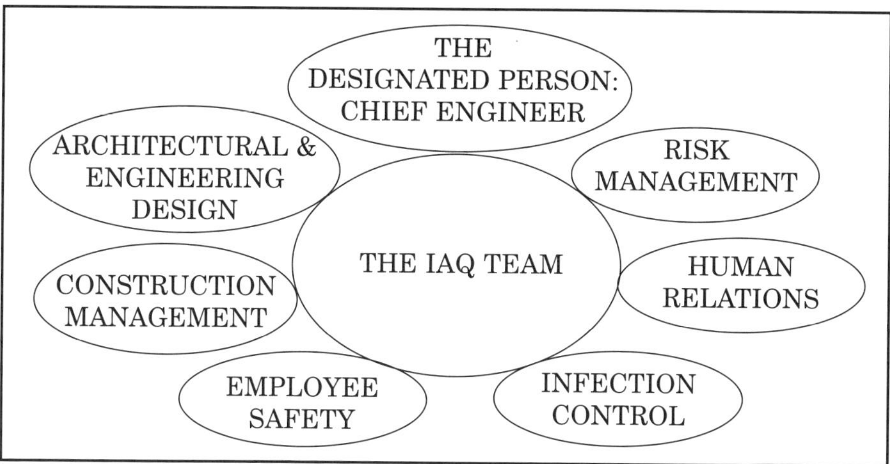

Figure 6-1: *Here is a possible format for an IAQ team.*

Source: Hansen W: Dealing with IAQ in the work place. Facility Management Journal *(May/Jun 1996).*

Implementing the IAQ Compliance Program

For the most part, health care organizations already have some type of IAQ compliance program in place. This section is merely a reiteration of the work practices and procedures already followed by most health care organizations. However, because of its importance, it bears repeating. Follow these steps if your organization has not already done so:

■ Ensure that all HVAC and exhaust systems are operating during all hours of occupancy. Of course, this element primarily refers to medical office buildings and other noncritical care patient areas that do not require 24-hour operation, where evening shut down of the HVAC system may occur. For evening operation for janitorial activities, boilers and chillers may be set back or shut down, but ventilation must remain active. Inspect internal conditions of the air-moving components of the HVAC system, which include the following: fans, coils, filters and filter boxes, mixing chambers, and ductwork. Note the conditions in a log, and take any remedial action indicated. (Examples of simplified log sheet can be found in Appendix C, page 140.)

■ Routinely inspect HVAC system mechanical equipment, including the automatic temperature control system for proper operation. These inspections should also verify that the fan rooms and plenum chambers have not been used to store liquids, combustible materials, or chemical products.

■ Ensure that health care workers are following good safety practices with regard to noxious or hazardous materials.

■ If some corrective action or renovation to the HVAC system is indicated in any of the inspections, have a follow-up implementation plan in place to address these needs.

■ Control and periodically monitor known indoor air contaminants. The relationship between contaminants and complaints is shown in Figure 6–2, page 69.

■ Keep relative humidity below 60%. This has been a very prominent issue in the south and southeast parts of the United States because of the mold and fungi problems. Humidity has an acute effect on a broad range of contaminant sources,[2] as shown in Figure 6–3, page 70. It should be noted that the surface humidity is a much more important factor for fungi and mite growth than the measured relative humidity in air.

One of the most crucial aspects of program implementation is the method of dealing with IAQ complaints from a human relations standpoint. This must be done in an unemotional and diplomatic manner, with care taken to establish a credible tracking system. When employees have complaints, listen to them—what they are saying and what they are not saying. Sometimes there are important clues for the astute facility engineer in what is not being mentioned. The team approach works best; bring the complainants into the problem-solving process.

There are four forms in Appendix D, page 141 to help guide this process, and they begin with the following steps:

1. *The complaint form.* Have the complainant fill out a complaint form in their own words to describe the nature of the complaint.

2. *The interview form.* After the IAQ team receives a complaint form,

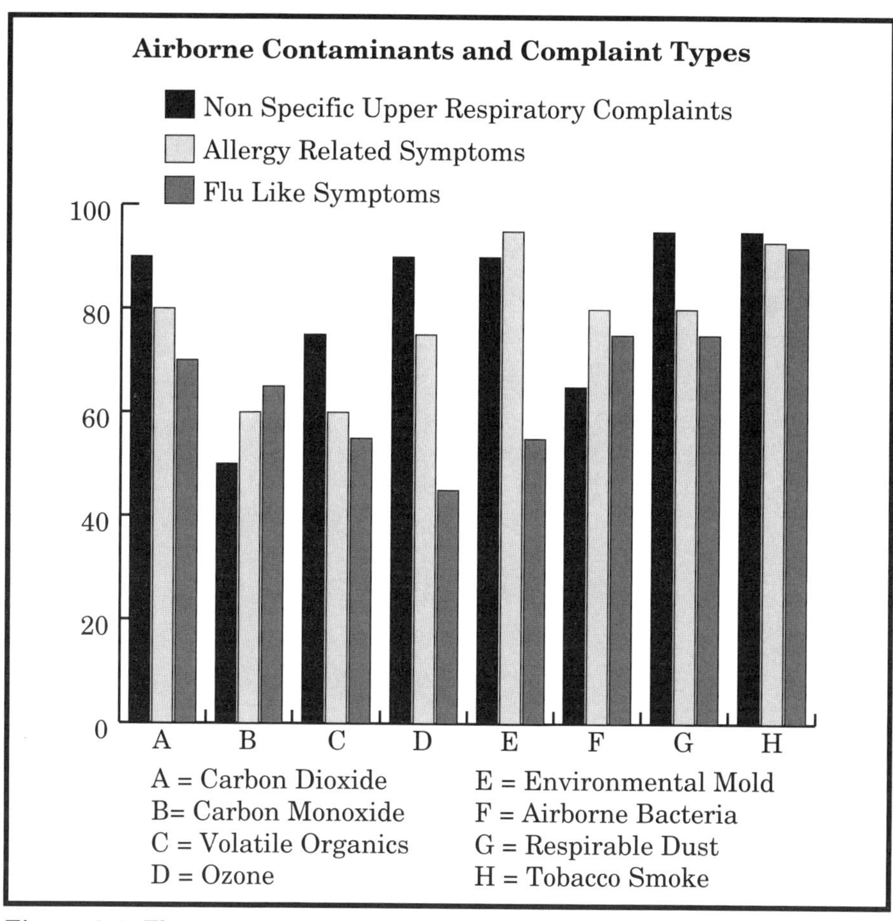

Figure 6–2: *The type of complaint varies with the type of airborne contaminant.*

Source: Hansen W: Dealing with IAQ in the work place. Facility Management Journal *(May/Jun 1996).*

they should follow it up with an interactive interview. This process could involve more than one interviewer from the IAQ team and more than one interview form. For example, the engineering department may do one, and based on this interview, it could be decided that a second interview with an infection control practitioner would be helpful. In some cases, it may be more useful for these two team members to conduct the interview together.

3. *The daily diary form.* In some cases, the complaint may be episodic in nature, particularly true in the case of odor related complaints. Giving

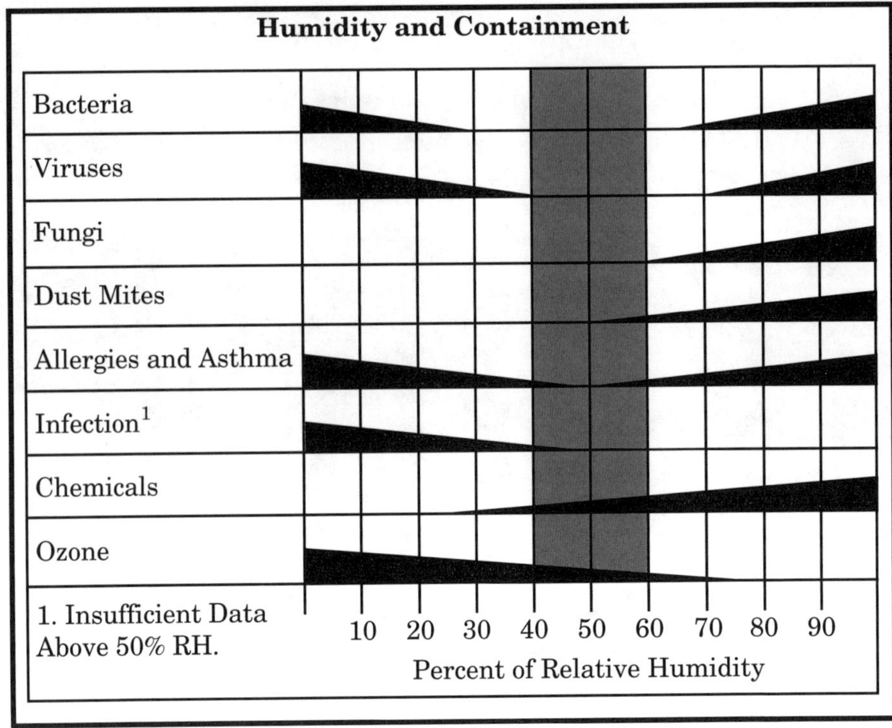

Figure 6–3: *The level of relative humidity affects the type of contaminants in the air.*

the person a diary form to keep at his or her work station or office helps narrow the area of investigation and makes the person part of the solution. Having a log to keep helps the person to be more objective about the complaint, and can provide some very useful data.

4. *The IAQ complaint log.* A complaint log is a summary tracking form for the facilities department. This one-line sheet acts as a "tickle file" so that the IAQ team can track the progress of each complaint.

Construction, Renovation, and Remodeling

It is a fact of life in any organization that some form of construction will take place after the facility is opened and occupied. This is particularly true in the health care industry, where there is constant pressure to update, expand, and modernize. Dealing with this while conducting the normal business of a health care organization is always a challenge.

70

In capsule form, it appears that we know what is required to prevent construction contamination,[3] but for unspecified reasons, these precautions are not being implemented. Several professional organizations, such as ASHRAE, have investigated the impact that construction activities have on IAQ and the unfortunate results from not taking adequate precaution. Chapters 2 and 4 touch on this issue, and Chapter 7 goes into more detail regarding this and other studies related to airborne contaminants generated by construction practice.

Employee Education and Training

IAQ is an often misunderstood topic, and the only way to gain a comfortable level of understanding when dealing with IAQ-related issues is through education and training. Health care facilities should provide basic education to all employees, not just those with responsibility for compliance and correction. The proposed OSHA regulations require education that includes the following:

■ Inform all employees of content of the OSHA standard and signs and symptoms of BRI and the requirement that employers evaluate building systems and take remedial measures if necessary when employee complaints about BRI are received.

■ The intent of this section is clear; *everyone* must know the basics and know that there is a standard and a facility plan for administrating it. General education on this issue is just as important for administrators and purchasing agents as it is for facilities and safety departments. Although the specific language in the proposed OSHA regulation refers to BRI, the intent is to address any IAQ related issue, because most employees cannot differentiate between a true case of BRI (see Chapter 2, page 15) and SBS (see Chapter 1, page 3).

■ Provide training for maintenance workers and building system maintenance and operation employees including:

• Use of personal protective equipment in operating and maintaining building systems (for example, respirators, dust masks, gloves, safety glasses and goggles);

- Maintaining adequate ventilation of air contaminants from cleaning and maintenance activities; and

- Minimizing adverse effects on IAQ during use and disposal of chemicals and other agents. (Special attention should be given to the patient room cleaning carts used by housekeeping to ensure that the cleaning and disinfecting chemicals are in covered containers when not in use.)

Although this section is directed toward the facilities and housekeeping departments, it also includes subcontracted services and vendors. As with any potential risk, it is the responsibility of the facility to ensure that service contractors and vendors are as aware of direct employees, and that they take the same precautions.

■ Make training materials available to employees and their representatives. At some point, OSHA representatives may visit the facility and will want to review the IAQ program, the implementation structure, and the type and extent of training and materials included. In this event, it is possible that these materials will have to be made available to the Director of the National Institute for Occupational Safety and Health in the US Department of Health and Human Services and the head of OSHA in the US Department of Labor.

All organizations experience some degree of turnover in their staffs, which implies that health care facilities must hold periodic training and education sessions for IAQ issues. Since IAQ programs are relatively new, it will be necessary to have follow-up training more often—to reinforce previous sessions and to present new material.

Record Keeping

In "Establishing the IAQ Program" section, I maintain that there are two distinct areas where record books are useful: one volume could deal with the definition of the IAQ program and employee issues and the other could deal with the facility and the mechanical issues. Record keeping could be done in loose leaf books for ease of use; the first loose leaf volume could be called the "Comprehensive IAQ Program Manual." In this, keep a copy of the program description, the IAQ team members, employee complaint forms, interview forms, and individual diaries.

The second volume could be titled "HVAC System Reports for IAQ Considerations." Here, keep a copy of the facility description, date built and dates of major remodels and additions with simple description of the work done, and a complete list of all HVAC equipment. This information is generally already available in the engineering office files and can be either relocated to this manual or simply copied to it. Also in this volume should be a copy of the IAQ-related inspection forms.

If the sample forms shown in the appendixes do not work into your existing HVAC system inspection and servicing log format, just use them as a guide for content, and find the most efficient method of integrating the items from these forms into the existing facility log format. The real key to designing a successful IAQ program is to make it as user-friendly to your specific organization as you can. This means understanding the *intent* of the IAQ program and implementing it accordingly, not attempting to comply chapter and verse with what someone else feels is a perfect program.

References

1. Environmental Protection Agency: *Building Air Quality, A Guide for Building Owners and Facility Managers,* EPA Document #EPA/400/1–91/033, and NIOSH Document #91–114. Washington, DC: Environmental Protection Agency and the National Institute for Occupational Safety and Health, Dec 1991.

2. Stepanek S: Better indoor air via HVAC operations: The quest and the quandary. *Buildings:* Mar 1992.

3. Kuehn TH: Construction/renovation influence on indoor air quality. *ASHRAE Journal:* Oct 1996.

Chapter 7: Recognizing IAQ Risk and Implementing an IAQ Program

Andrew J. Streifel, MPH, REHS

Mr Streifel is Environmentalist, Department of Environmental Health & Safety, University of Minnesota, Minneapolis. He has more than 20 years of experience as the Hospital Environment Specialist for the Department of Environmental Health and Safety at the University of Minnesota and as adjunct faculty in the School of Public Health. Mr Streifel has been widely published in professional journals and in newspaper articles regarding his experience in more than 150 health care facilities relating to indoor air quality issues.

The evolution of health care, with modern therapeutic and diagnostic tools, requires that health care facilities recognize hazards for patients and employees in the spectrum of health care delivery. Personnel working in the acute care hospital environment should be aware of airborne hazards that must be controlled in order to provide beneficial patient treatment. The hospital environment must provide safety for employees, patients, and visitors from physical, biological, and chemical agents found in the acute patient care setting. Not recognizing and controlling such hazards can lead to health concerns for employees, additional health issues for the patient, and costly renovations or lawsuits for hospital management.

This chapter outlines the process for implementing a proactive indoor air quality (IAQ) program:

■ Conducting a risk assessment;

■ Managing airborne spread infectious agents;

■ Establishing control measures for safe IAQ; and

■ Monitoring chemical air quality.

Following this process can prevent or minimize airborne hazards in the environment of care.

Conduct a Risk Assessment

Air quality can become an issue during treatment of specific diseases that render patients susceptible to opportunistic infection from airborne infectious agents (see Chapter 4, page 44). Medical intervention of certain diseases causes the patient to become extremely susceptible to common environmental microbes which have only been recognized as hazardous within the past 30 years. Other hazards may be associated with occupational exposure to medicated aerosols or sterilant fumes used to reduce the risk of infectious disease from devices utilized to diagnose and treat the patient (see Chapter 5, page 57). For example, irritating airborne fumes and/or airborne infectious agents may be associated with endoscopy procedures used to diagnose pulmonary diseases. During surgery the patient is administered conscious-altering gases and the waste gas can affect exposed personnel.

To determine where resources must be allocated for reducing the risk from health care airborne hazards, health care facilities must assess the patient and employee risk. Due to the nature of some treatments for malignant disorders or organ transplants, some patients are at risk for infection—and even death—from common environmental microbes after such treatment. The airborne route of hazardous agent transport must be recognized for relative risk to the patient and employee. Recognizing that the patient is administered medications that may cause a potentially hazardous occupational exposure to the health care worker is essential. Safety in the environment of care for the patient and employee is a primary objective.

In today's health care facilities recognizing potential hazards is of primary concern for focusing resources.[1] Patient susceptibility is variable and the highest priority must be associated with factors that could be life threatening. Cancer therapy, for example, can make patients extremely susceptible to infection from common airborne fungi. Such infections are not amenable to therapy; hence, they must be prevented. Likewise, during surgery if waste anesthetic gases are not controlled, employees may suffer short- or long-term effects. In considering how much risk to allow, health care facilities should take into account these factors:

■ *Patient susceptibility:*

- Bone marrow transplantation;
- Cancer immunosuppression;
- Solid organ transplantation;
- Premature birth babies;
- AIDS patients; and
- Patients compromised by certain interventions, such as surgery.

■ *Employee susceptibility:*

- Immunocompromised employees;
- Allergic employees; and
- Employees working with hazardous materials.

■ *Chemical or physical hazards:*

- Medicated aerosols;
- Vapor or fume; and
- Smoke plume.

■ *Biological agents:*

- Patient generated airborne infectious agents; and
- Environmental opportunistic infectious agents.

The airborne spores from indoor sources, such as fungal growth or spore accumulation, could cause allergic reactions in susceptible employees. The same spores can cause death in bone marrow transplant patients. The diverse nature of the hospital environment requires broad insight to assess the risk factors of airborne agents. The mold growth resulting from prolonged water damage should be recognized as a potential cause for the range of hazards, from mild irritation to loss of life, to the respective hospital occupant.

Safe IAQ means anticipating the presence of airborne biological and volatile chemical hazards. The modern health care environment should emphasize cost-effective measures that recognize risk and allocate safety measures to prevent such hazardous IAQ exposures.

Manage Airborne-Spread Infectious Agents

Airborne-spread infectious agents are relatively few in comparison to the total number of microbes available to infect humans. Still, we must be aware of those from the environment, as well as those from the patient. When a patient source of airborne microbes is detected, the patient must be isolated for the protection of all health care facility occupants due to the airborne transmission potential. Delayed diagnosis of these diseases is often the primary reason for occupational exposure to patient-derived airborne infectious diseases in the health care setting.

According to the isolation guidelines enumerated in *Infection Control and Hospital Epidemiology,*[2] airborne spread infectious diseases include: Tuberculosis, chicken pox, and measles. Patients with these infectious diseases require isolation, including a sealed room, increased room air exchanges, exhaust ventilation, and airflow into the room (negative pressure). Such special ventilation rooms must be monitored to provide ventilation assurance during airborne infectious disease isolation.

Environmental organisms. Disease treatments can cause patients to become ultra-susceptible to environmental opportunistic infectious agents causing aspergillosis or legionellosis. The hospital is required to provide protective measures to prevent exposure to environmental infectious microbes. Airborne infection from opportunistic fungi is greatest for oncology treatments causing neutropenia. Treatments that change the immune system during transplantation of either bone marrow or a solid organ are especially critical. It has been recognized that construction and renovation projects also provide a significant risk factor for immunocompromised patients.[3]

Hospital building management must recognize which environmental conditions cause the extreme risks; these conditions often create IAQ problems for health care workers exposed to airborne fungal agents as well. While fungal contamination affects both the employee and patients, the latter have a much lower tolerance for exposure.

As is discussed in previous chapters, deferred maintenance of the building and building systems is often at fault for creating sources of environmental contamination. Failure to correct moisture problems and not routinely evaluating ventilation parameters in critical patient care areas significantly contributes to occupant risk.

Airborne biological agents. The Centers for Disease Control's (CDC) list of known airborne spread microbes[2] is not large; however, the known agents causing tuberculosis, measles, and chicken pox represent considerable risk; for example, tuberculosis worldwide is a serious infectious disease and in hospitals represents an occupational risk. The CDC's list of airborne microbial agents follows:

■ ***Patient infectious:***

- TB (tuberculosis);

- Varicella (chicken pox); and

- Rubeola (measles).

■ ***Environmental infectious:***

- *Legionella* genus bacteria, primarily *L. pneumophila* (serogroup II);

- Airborne fungi *(Aspergillus fumigatus, A. flavus, A. terreus, Fusarium* genus, and other airborne fungi capable of growth at 37°C).

■ ***Environmental sensitizing:***

- A variety of filamentous spore forming agents affect occupants due to allergic response;

- Toxin producing microbes;

- Insect debris; and

- Latex.

The hospital filtration systems if properly designed will remove more than 98% of spores >1.0mm in diameter,[4] provided that the systems are properly installed and maintained. Hospitals must also recognize the fact that mold proliferation indoors is often caused by uncontrolled water sources. Water leakage often becomes a deferred maintenance issue and is not factored in as a cause for IAQ hazards.

Medicated aerosols. Therapeutic agents given to patients can cause them to cough and release infectious bacteria, especially *Mycobacterium* *tuberculosis.* Containment of medicated aerosols, such as pentamidine, is necessary during administration to prevent occupational exposure. Therefore, a protective measures standard for working with airborne particulates must be implemented.

Volatile chemicals. Anesthetic gases are administered in surgery as well as in outpatient procedures such as dental surgery. Medical personnel will be exposed to the waste gas unless they are properly scavenged from the work area. Outpatient surgery often involves different methods for administering anesthetic gases, which can affect personnel exposure to the waste gas. Whereas inpatients often are intubated for controlling the gas, outpatients are administered gases through masks, which do not contain the gas as readily as intubation. The threshold limit value is established for a variety of anesthetizing agents as shown in Table 7–1, below.[5] To date, no regulatory agency has recommended a permissible exposure or ceiling limit.

Equipment maintenance
    Tubing connections
    Dome vale or bellows leakage

Procedure practice
    Mask versus intubation gas administration
    Pediatric cases
    No waste gas scavenging

High pressure gas
    Machine connections
    Loose valves

Vaporizer
    Leaking vessel seal
    Spills

Table 7–1: *Waste Anesthetic Gas Problems*

Preventive maintenance and management of scavenging and building ventilation are essential to minimize exposure to these gaseous agents. Employee practice should be monitored in order to assess minimal exposure.

Disinfectants and sterilants. When agents such as gluteraldehye and formaldehyde are used, an exposure level should be documented (see

Table 7–2, below). It is imperative to monitor personnel exposure to these agents and provide environmental control to ensure employee safety. Containment methods have been prescribed,[5] and such hazard control should include sufficient ventilation for dilution and capture of the airborne disinfectant, sterilant, and fixative agents. Monitoring ventilation parameters and an employee's breathing zone with dosimeters provides documentation for quality management.

AGENT	**PPM**
Gluteraldehyde	0.2 ceiling limit
Formaldehyde[1]	0.1 TWA
Peroxide products (peracetic acid)	0.1 TWA
Ethylene oxide	1.0 TWA

[1] Monitoring required by OSHA
TWA = Time-weighted average

Table 7-2: *Agents With Low Exposure Limits*

Laboratory chemicals and biological agents. Common airborne agents in laboratories are generally contained in fume hoods and biological safety cabinets.[6,7] To contain microbes during use of aerosol generating devices, such as decontamination burners or centrifuges, containment hoods should also be used.

Those performing and assisting autopsies risk exposure to infectious or chemical agents. The use of high-speed saws can release significant aerosol of infectious agents such as *M tuberculosis*. Bloodborne pathogens, such as HIV or hepatitis viruses, have also been thought to be transmitted via airborne exposure, though such theories have not been substantiated.

Laboratory ventilation control and training are necessary for effective containment of airborne hazards. Because a laboratory is expected to have negative pressure air flow, it becomes a sink for other airborne contamination. Defining control parameters for monitoring is necessary to ensure minimal personnel exposure, including laboratory ventilation, containment hood certification, and employee training for safety management.

Intake exhaust fumes. Intake exhaust fumes is covered in previous chapters but must be reiterated because many indoor air complaints come from vehicle or generator exhaust. Building systems for preventing such nuisance and anxiety-producing odors are cost-effective when considered in the original design; after-the-fact corrections can be very costly.

Exhaust fume management may include recirculating affected ventilation with minimal outside air, charcoal/absorbent filtration, and traffic and construction management. Engine idling creates annoying fumes, affecting personnel. Signage in obvious locations will help to notify drivers to shut off engines, and planning helicopter landings away from intakes will help. If, however, the heliport is near critical building air intakes, hospital management should consider a policy to reduce outside air intake or guidelines requiring helicopters to shut down.

Parking lot resurfacing and roofing odors are also a source of many complaints. Prevention may be impossible but awareness and complaint response help to assure occupants that the chemicals they are smelling are not at hazardous levels. Employees and patients prefer response and assurance that the source is known and exposure may be short term, rather than anxiety and anger that often develop from being ignored as a "nuisance complaint."

Establish Control Measures for Good IAQ

As noted in Chapter 1, the efforts to control exposures from airborne hazard should be delineated and should begin with source management. Source management for environmental microbes may be more definable than, for example, determining who has infective tuberculosis.

Patient protection. "The Guidelines for Prevention of Nosocomial Pneumonia, 1994"[8] provides parameters to help prevent nosocomial aspergillosis and legionellosis. Airborne fungal control is required for patients undergoing severe immunosuppression during treatment for certain types of cancer, solid organ failure, or specific immune deficiency diseases. The CDC recommends enhanced ventilation with increased room air exchanges, high-efficiency filtration (99.97% at 0.3mm sized particles), positive pressure, and construction management.[8]

Because it is difficult to treat patients after an infection is caused by airborne fungi, such as *Aspergillus fumigatus,* prevention must be a part of procedural practice. Recognizing the deficient conditions that cause airborne infection to occur requires experience in building IAQ investigation. For example, if airborne fungi are found during air sample analysis in the lobby in relatively high numbers and also found in a bone marrow transplant unit (BMTU), one must investigate the air balance between the two areas. In such an instance, the pressure relationship between the lobby and the BMTU will pull airborne spores from the lobby to the sensitive area. To correct this, the BMTU and adjacent areas must be rebalanced to ensure that air flows from cleanest to less clean; but, manipulating an air handling system to correct airflow may also release infectious spores in a contaminated ventilation system.

Control of hospital ventilation systems is complex and institutions providing medical service for oncology and solid organ transplant require active air quality management. Unfortunately, the original design of the building often emphasizes ventilation energy management at the expense of infection control ventilation. For example, the corridor and nursing support areas in modern hospitals are placed on a variable air volume ventilation system, whereas the patient rooms are constant volume. Due to close tolerance and little design guidance, it becomes difficult to ensure a constant pressure differential between critical spaces. These conditions have been observed to cause a reversal of airflow in surgery, BMTU, and airborne infection isolation rooms. To avoid this, the design criteria for ventilation of the patient care areas must be communicated during design and followed through during occupancy with verifiable parameters, such as the following:

■ Positive pressure room—greater supply than exhaust return air volume;

■ High efficiency filtration—point of use filters in the patient rooms, 99.97% efficient @ 0.3mm;

■ Increased room air exchanges—more than 12 per hour;

■ Clean to dirty airflow—patient to employee areas; and

■ Self-closing doors—a well-sealed room.

To verify that the ventilation control design is correct, particle counters both viable and nonviable—can be installed to determine air quality content. Facilities management can also install pressure gauges to read air balance data are in special ventilation rooms. It is essential to collect these data at the commissioning of the space—before occupancy, as well as after—so that the data can serve as background for comparison purposes. Should litigation result from building occupants who have infection from environmental microbes, it is critical to determine if community standards as prescribed by the American Institute of Architects and the CDC were followed. Defining disease as nosocomial versus community-acquired is essential to avoid confusion. The responsibility of disease recognition rests with the health care facility epidemiology/infection control personnel. This group should help recognize the respective high-risk areas.

Airborne infection isolation. The presence of an infectious patient is hazardous to a susceptible employee. The airborne-infectious agents causing tuberculosis, chicken pox, and measles must be contained in special ventilation airborne infection isolation areas that have the following elements:

■ Exhaust to the outside or, if recirculated air must pass through, 99.97% efficient HEPA filtration;

■ Sealed rooms to prevent air infiltration, which could cause reversal of airflow;

■ Self-closing doors;

■ Greater exhaust than supply air volume by 50 cfm;

■ Ante rooms where appropriate;

■ Increased room air exchanges greater than 12 per hour; and

■ Capable of more than 0.001" water gauge (w.g.) pressure differential.

Containing patient airborne infectious disease is a priority, and health care facilities must assess risk with respect to community disease through involvement with infection control and local health agencies. Such interaction will help establish the degree of involvement recommended by the CDC[9] in assessing where precautions must be taken, as shown in the following list:

■ **Positive pressure rooms (protective environments):**

- Operating rooms;
- Bone marrow transplant;
- Oncology services; and
- Solid organ transplant services.

■ **Negative pressure rooms for airborne infection isolation (AII):**

- AII patient care rooms;
- AII suspect emergency rooms; and
- Procedure room for bronchoscopy or medicated aerosol administration.

Operating rooms. Operating rooms present a difficult issue when airborne infectious agents are considered. For example, conditions in the HVAC system can develop due to minimal filtration or excess moisture causing mold accumulation. Or a patient could have an extra pulmonary tuberculosis lesion to be removed, which may place the surgical team at risk. The control of the protective operating room environment should include source management for gases, plumes, and patients. Reduced ventilation due to negligent maintenance can create risk to the surgical staff because the ventilation does not reduce the exposure to the anesthetic gas.

Monitoring ventilation parameters. Both the protective environment and airborne infection isolation rooms should be monitored. The monitoring devices must be easy to read and interpret. For example, the smoke stick that generates a visible plume to observe airflow movement demonstrates direction and relative pressure. A pressure monitoring device will demonstrate pressure differentials; mechanical and nonmechanical pressure monitors are available. Nonmechanical flutter strips or "ping pong ball" indicators are simple and effective for demonstrating airflow direction. A pressure differential for special ventilation rooms should be consistent and not subject to building pressure changes. Ideally, room pressure should be at or above 0.01" w.g. or 2.5 pascals. Rooms with changeable ventilation are discouraged. Utilizing ante rooms for special considerations, such as bone marrow transplant rooms, for isolating airborne infection is helpful.

Periodically, facilities management staff should evaluate the special ventilation areas for air balance and room air exchanges and check the airborne infection isolation rooms monthly according to the tuberculosis isolation guidelines. Although the positive pressure protective rooms have no guidance for room parameter evaluation, measurements of the special ventilation rooms should be conducted at least quarterly. When filtration measurements are taken before occupancy, these baseline data become the benchmark for future problem solving or quality management for ensuring the safety of occupants in the environment of care. Filtration checks using particle counts of both viable fungi and nonviable particles, along with air balance and pressure information, should serve as baseline information.

Monitor Chemical Air Quality

When chemical sources are known in the operating rooms, central sterile processing, or laboratories, there should be a chemical hygiene plan in place that includes protocols for the safe handling and storage of the variety of chemicals present in the hospital. The following is a list of suggested protocols:

■ Laboratories should monitor:
  • Volatile chemicals;
  • Acids;
  • Formaldehyde (OSHA required); and
  • Gluteraldehyde.

■ Central processing areas should monitor:
  • Gluteraldehyde and
  • Ethylene oxide.

■ Operating rooms should monitor:

  • Waste anesthetic gases and

  • Methyl methacrylate.

When dealing with the chemicals that have ceiling limits or threshold limit values, exposure levels should be documented for ensuring safe environments using dosimeters or real-time monitors (for example, infrared analyzers) to determine the exposure limits. If, for example, complaints have been recorded in the endoscope processing area involving exposure to gluteraldehyde disinfectants, measurements will determine the levels of exposure. The measures to reduce that exposure can be verified with before- and after-adjustment exposure monitoring.

Operating rooms. Employees can be exposed to the waste anesthetic gas in operating rooms or labor and delivery areas to varying degrees depending on the method of gas administration, scavenging capability, equipment maintenance programs, and whether local ventilation is adequate. Routine preventive maintenance of the anesthesia equipment to ensure minimal leaks around high pressure connections and vaporizer seals is essential, as is maintenance of hoses, valves, and scavenging systems, to minimize exposure. Cleaning up liquid anesthetics spilled while pouring is an especially serious problem. Training employees to handle anesthetics, emphasizing exposure reduction, is mandatory to reduce those exposures. Short-term exposures can create behavior changes, including irritability and judgment alterations, and long-term exposure can affect reproductive organs all of which should be eliminated through exposure control.

Other areas where hazardous gases or chemicals are used should have monitoring devices as well. For example, chiller and refrigeration rooms where chlorofluorocarbons are present should have oxygen detectors; if such chemicals are spilled, oxygen levels can drop below safe limits. Ethylene oxide monitors should be in place to detect leaks or equipment malfunction in areas such as central processing. (Note that certain areas should have continuous monitoring equipment installed while other areas should be evaluated periodically.)

Construction management. The most obvious disruption of the indoor environment is often overlooked as a "normal" occurrence. Utility upgrade and patient service changes usually mean that the health care facility is in a state of continual construction or renovation. With construction, the most critical consideration is preventing release of hazardous material and disruption of critical ventilation. The recent AIA design and construction guides for hospitals and health care facilities show the importance of construction management by dedicating an entire chapter to it. The following are some important points from that chapter:

■ Planning and design:

- Risk assessment to determine critical patient areas;

- Safety and infection control involvement;

- Design criteria for air quality control; and

- Airflow management during construction/renovation.

■ Phasing:

- Ensuring clean to dirty airflow;

- Roof protection from puncture;

- Protection of air intakes from exhaust;

- Ventilation interruption notification; and

- Water damage management.

■ Commissioning:

- Specifying acceptance criteria for IAQ;

- Verifying filtration and air balance specifications;

- Special ventilation areas for surgical services, protective environments, airborne infection isolation;

- Laboratories and local exhaust systems for hazard control ventilation; and

■ Deficiencies are unacceptable.

Construction is associated as a risk factor for opportunistic airborne fungal infection in immunocompromised patients because normal ventilation is disrupted, possibly releasing hazardous airborne spores into the hospital environment. The prevention of infection or other IAQ problems associated with construction requires careful management in the environment of care. Complaint-related issues are often associated with vapors, fumes, or other odors released during construction. While most health care facilities recognize that implementing interim life safety measures (ILSM) is essential during construction, some may have overlooked the critical factor of occupant protection from hazards other than fire.

Enlightened health care organizations faced with disease and complaints associated with construction disruption are adjusting the planning, implementation, and inspection process for management of renovation and construction to include infection control and IAQ items. This means that the construction risk factors are identified, especially those relating to the at-risk populations, and which phase of the project might create risk. The phases that usually create the greatest risk include: window or wall removal; ventilation outages; application of volatile chemicals; and placement of combustion engines. Each project must include methods to control aerosols and volatile chemicals as well as the inspection process to prevent incident.

Water damage from construction should be anticipated as a standard occurrence; but without appropriate reaction on the part of facilities management, standing water can create conditions for ubiquitous mold to proliferate and contaminate the IAQ of the local environment. Too often leaks and spills (for example, from roof puncture, plumbing testing, frozen pipes, and condensation) are ignored, allowing prolific spore production, which creates odors and when disturbed, airborne allergic and infectious particles. At room temperature, the presence of water with a cellulose nutrition source creates the conditions for mold growth within 72 hours. To prevent such situations, construction teams must be vigilant and remove water-damaged material if it cannot be dried in less than 48 hours.

The obvious is not always acted upon during the process involved with acceptance of new or renovated hospital space. For example, assurance of ventilation according to design is basic to all indoor environments; but the pressure to accept the building and begin the intended function is overwhelming. Under such conditions, ventilation and other building factors are often accepted with deficiencies. Such deficiencies, especially with respect to the ventilation for critical space, can create life-threatening conditions or discomfort factors that are extremely expensive to correct after occupancy. Such factors are common enough to warrant the AIA to refer to construction factors for planning, phasing, and commissioning hospital construction projects in regulatory language.[10]

Coordinating project outages, water damage management, roof protection, and pest management might cause the contractors additional expense. But dealing with such matters expediently can actually help the project progress on schedule because the defined process helps those in charge make decisions. For example, a hypothetical hospital retains consultants to help decide IAQ issues during construction. During a major water damage incident in a newly completed operating room area, the consultants recommend immediately evaluating the area with a wet test meter to help define the water damaged areas for removal. The water-damaged material is removed in less than 48 hours, and the project can proceed because the growth potential is obviated. The project continues in spite of the fact that several hundred feet of duct insulation and dry wall were removed. Such action would prevent the owner and contractor from entering into a dispute, which could have resulted in delays for deciding what to do, during which time mold could grow.

References

1. U.S. Department of Health and Human Services: *Guidelines for Protecting the Safety and Health of Health Care Workers.* Washington, DC: U.S. Government Printing Office, 1988.

2. Garner JS: Guideline for isolation precautions in hospitals. *Inf Cont and Hosp Epid* 17:53, 1995.

3. Kuehn TH, Gacek B, Yang C-H, et al: *Identification of contami-*

nants, exposures, effects, and control options for construction/renovation activities. ASHRAE Trans 102:89–101, Nov 1996.

4. OSHA: Controlling occupational exposure to hazardous drugs. *In Directorate of Technical Support.* OSHA, 1995.

5. Committee on Industrial Ventilation: *Industrial Ventilation, 20th Edition: A Manual of Recommended Practice.* Cincinnati: American Conference of Governmental Industrial Hygienists, 1988.

6. U.S. Department of Health and Human Services: *Biosafety in Microbiological and Biomedical Laboratories.* Washington, DC: U.S. Government Printing Office, 1993.

7. Committee on Prudent Practices for Handling, Storage, and Disposal of Chemicals in Laboratories: *Prudent Practices in the Laboratory, Handling and Disposal of Chemicals.* Washington, DC: National Academy Press, 1995.

8. Tablan OC, Anderson LJ, Arden NH, et al: Guideline for prevention of nosocomial pneumonia. *AJIC Am J Infect Control* 22:247–92, 1994.

9. APIC: *Infection Control and Applied Epidemiology Principles and Practice.* St. Louis: Mosby.

10. American Institute of Architects: *1996–97 Guidelines for Design and Construction of Hospital and Health Care Facilitites.* Washington, DC: AIA Press, 1996.

Chapter 8: Economic Considerations and the Benefits of an IAQ Program

Stephen M. Burch, PE, and Brian Krafthefer, PE, DEE

Stephen M. Burch, PE, is Principal Engineering Consultant, Honeywell Inc, Health Care Business Unit, Nashville. Mr Burch is a registered engineer with more than 16 years of design and construction experience in the health care field. He is an active member of the American Hospital Association and the American Society for Healthcare Engineering.

Brian Krafthefer, PE, DEE is Staff Research Engineer, Honeywell Technology Center, Minneapolis, Minnesota. Mr Krafthefer is a registered engineer with an educational background in mechanical engineering and physics. He has more than 18 years of experience in the research and application of HVAC systems and IAQ.

The effects of poor IAQ cost the U.S. economy billions of dollars annually. The Environmental Protection Agency (EPA) estimates losses in excess of $40 billion per year in worker productivity and $1.2 billion in annual medical costs.[1] Legal liability, insurance, property damage, and mitigation/remediation efforts add to the economic impact. The magnitude of these costs underlines the importance of economics in the discussion of IAQ. Economics are responsible for most IAQ problems, and they influence the decision-making process, as the cost of providing a high-quality indoor environment is weighed against the costs of IAQ problems. The cost of energy, maintenance, and operations required to maintain a quality environment, however, is small compared to "people costs" and the effects of the indoor environment on their performance and well-being.[2]

Definition of Costs

The economic impact of IAQ on the hospital can be divided into two basic cost categories:

1. ***Compliance costs*** are those costs incurred by the hospital to ensure that the air quality within the facility is maintained at a proper level. Compliance costs are typically "direct costs" in that they stem immediately from and are defined by a close causal relationship with IAQ issues. These costs include HVAC/plant energy, equipment maintenance, repair and replacement, IAQ programs, and problem mitigation and remediation.

2. ***Avoided costs*** are "indirect costs" that trace their root cause to IAQ issues but are not immediately identifiable with these issues; they are costs for which the facility is at risk should an IAQ problem develop. They can be incurred by the health care industry as a whole and are often passed on to the consumer or general public.

Although there are little hard data available for estimating the economic impact IAQ has on health care, the EPA has made an estimate of poor IAQ costs in its Report to Congress on IAQ.[3] The EPA report addresses three major categories: materials and equipment damage, direct medical costs, and lost productivity. There were many other costs that were not considered in the study, such as legal liabilities, workers' compensation and insurance costs, medical and patient care costs, and property damage.

IAQ Market Overview

Economic history. As shown in Chapter 1, the emergence of IAQ as a documented health concern can be traced to the energy crunch of the early 1970s.[4] Energy efficiency took precedence over ventilation and ultimately became a significant factor in the growth of IAQ problems and SBS.

The results of a National Institute for Occupational Safety and Health study (shown in Table 8–1, page 95) identified ventilation as the primary cause of IAQ complaints in 52% of the cases.[5] Additional research has shown that inadequate ventilation is usually the result of energy management strategies or a lack of proper maintenance,[2] both of which are driven by economic decisions.

Causative Factor	Percentage of Cases Affected
Inadequate Ventilation	52%
Contamination Inside the Building	15%
Contamination Outside the Building	10%
Microbial Contamination	5%
Contamination from Fabrics	4%
Other Sources	13%

Table 8–1: *Causes of Indoor Air Quality Problems*

Market forces. The health care industry is beginning to realize that IAQ improvements make economic sense, and voluntary efforts to improve IAQ are increasing. Despite these trends, the effects of voluntary efforts have not been sufficient to eliminate the need for regulatory action.

Regulatory activity. Economic theory suggests that the need for regulatory action is reduced where market forces work to ensure worker safety and protection. This, however, has not been the case in the private sector today. Employees are not always knowledgeable of their exposure to hazards or their risk of injury; and employers are not always willing or able to provide adequate information. True mobility between jobs and employers is not a reality. In many cases, workers are forced to choose among a good job in a hazardous environment, a low-paying job in a safer environment, or unemployment. Furthermore, employers do not solely bear the cost of injury or illness; occupational health and safety costs are passed on to the general public through taxes and social programs, further reducing their economic incentive.[6]

Other remedies. Tort litigation and workers' compensation claims are becoming increasingly common in IAQ cases as forms of worker protection. Tort liability does not totally internalize the cost, and those responsible are often able to transfer their risk to insurance companies, again reducing the economic incentive for corrective action. Furthermore, the tort system is not applicable in cases between employers and employees in which worker compensation is the only remedy but is typically inadequate in cases of occupational injury and illness. Here again, the public bears a large portion of the financial burden, and employers have a greater incentive to reduce premiums by contesting claims than to implement programs to correct deficiencies.[6]

Market forces have proven ineffective in providing adequate levels of safety, and tort liability and workers' compensation have provided insufficient protection.[6] Hospitals also need to protect the patient from any indoor hazards. As shown in the Introduction of this book, OSHA, NIOSH, ASHRAE, and other organizations have or are in the process of addressing this issue.

Projected costs. For the health care industry, OSHA estimates an annualized cost of $496.3 million for compliance with proposed IAQ regulations. This number represents a total expenditure of 1.28% of annual profit and 0.08% of annual revenue.[6] In 1992, there were 6,509 hospitals in the United States, with a total of 1,174,000 beds that had overall operating expenses of $282.5 billion.[7] Payroll accounted for $108.5 billion for the 4,466,000 employees at an average hourly wage of $13.47. OSHA estimates that 30% have IAQ problems, resulting in exposure to 1.3 million employees.[6]

Compliance costs. The cost to maintain a quality indoor environment and to comply with proposed IAQ regulations can be separated into four basic categories: administration, equipment, operations and maintenance, and energy.

Administrative costs. The elements of an IAQ program and the procedures for developing and implementing such a program are covered in Chapter 6. The cost to develop an IAQ program will vary depending upon the facility and the complexity of its systems. It is suggested that hospitals also consider establishing a training budget based on a minimum, one half-hour training program for initial orientation of all affected employees, with additional sessions for new employees and updates as required.

Equipment costs. The cost of IAQ can also include substantial equipment upgrade, retrofit, repair, and replacement. Improving existing systems carries a significant energy penalty as air flow is increased. The impact of increasing the air flow of an existing fan is expressed by the fan law as follows:

$$(CFM_2/CFM_1)^3 = (BHP_2/BHP_1)$$
$$\text{where: } CFM = \text{air volume in cubic feet per minute}$$
$$BHP = \text{fan brake horsepower}$$

A 10% increase in CFM increases fan horsepower and energy consumption by 33%, typically requiring a larger fan motor, with a possible impact on electrical service, cooling capacity, and duct size.

In many cases, it may be cost effective to install a comprehensive building control and energy management system to monitor the equipment and ensure that it is functioning properly. It also gives facility managers the opportunity to implement control and energy savings strategies while maintaining good IAQ.

One-time IAQ equipment improvement costs range from $1.14 per square foot[6] to $1.50 per square foot.[8] Eto and Meyer estimated the cost to increase the outside air rate from 5 CFM to 20 CFM/ person at $0.19 to $0.32 per square foot.[9]

Operations and maintenance costs. Inadequate maintenance and housekeeping procedures have been identified throughout this book as major factors in IAQ problems. The cost for each health care facility will depend on a number of factors including the current maintenance program in place and the expertise of the staff. OSHA estimates the cost of IAQ-related maintenance at $.17–$.25 per square foot per year.[6] When a comprehensive program is in place, there may be no additional cost involved to meet IAQ maintenance standards.

It may be necessary to complete a test and balance program to correct existing deficiencies and bring equipment and systems back to proper operating conditions. Coupled with the balance is a thorough cleaning of the HVAC system: air handling unit interior surfaces, coils, drain pans, plenums, and ductwork. This procedure can improve IAQ and help pay for additional IAQ improvements through energy savings, which result from better heat transfer through the coil and less static pressure drop across the coil.

Energy. Total energy cost for a typical hospital averages between $2 and $5 per square foot per year.[8] A study conducted by California's Office of Statewide Health Planning and Development found an average cost of $2.81 per square foot. Another study conducted by a private corporation shows for the average hospital that HVAC related equipment accounts for approximately 56% of the annual electrical energy usage,[10] a significant amount of which is directly related to IAQ.

Several studies have been done to quantify the effect of outside air rates on annual energy costs. One study of a commercial building, such as a medical office building, where recirculation was permitted showed that increasing the outside air rate from 5 CFM/person to 20 CFM/person increased the energy cost 4.5%[9]. A similar study[11] involved hospitals that allowed some areas of recirculation. In this case, the ventilation rate was increased from 5 to 25 CFM/person and showed a corresponding energy cost increase of 2.7% to 3.5%. In full-service hospitals where large quantities of outside air are necessary, energy savings can be achieved through the use of heat recovery units.[4]

The cost of energy is typically about 4% of the total hospital operating budget.[9] A 10% increase in energy cost is only a 0.4% increase in overall operating cost; an amount substantially less than the economic impact of poor IAQ.[1] Even small improvements in productivity more than pay for reasonable increases in ventilation.[12]

Energy Savings and Productivity. Studies have been performed on energy and productivity dealing with the impact of IAQ on cost for commercial buildings.[2,8,13,14] If the energy cost increases generated from increasing the outside air percentages shown previously (cost) is compared to the increases in productivity (benefit) identified in these studies, the cost/benefit ratio ranges from 10:1 to 15:1.[2,9,11] One study[15] also examines the energy and cost implications for ASHRAE Standard 62–1989 but is limited to administrative areas only in hospitals. The health care environment has internal zones that are unlike most commercial office buildings for which these studies are conducted.

The basis for the energy usage, furnishings (standard office variety), and cost basis for administration were taken from Woods' study.[12] Because nonadministrative areas have such a high capital equipment cost and associated technical cost, there is even a larger ratio of operating-to-energy costs than is found in most commercial buildings.

One of the strongest measures of productivity has been absenteeism.[2] Absenteeism can pose multiple costs for the employer through decreased productivity, decreased morale, increased health costs, and potential increase in workers' compensation claims and union grievances.

Improved ventilation can often yield net savings,[16] and increased ventilation rates will decrease absenteeism. One study[17] shows that for a 100,000 square foot building with 667 employees, an increase in ventilation from 5 to 20 CFM/person would cost an estimated $5,575.70 per year. The estimated first costs for improving the system were $28,500. For an average 2.6% absenteeism rate among white collar workers, 50% of the lost days are due to respiratory problems.[18] In poorly ventilated buildings, up to 33% of those lost days may be associated with inadequate ventilation. The study[17] indicates that, for this situation, a first-year savings of $70,235 is due to a decrease in the absenteeism produced by increasing the ventilation of the building, with a savings of about $65,000 for each of the following years. It has been stated that a 1% increase in worker productivity justifies a 15% increase in the building construction or a 100% increase in the operating costs.[19]

Legal Costs. The impact of legal action is becoming greater for hospitals with the lack of standards or regulations. Because such standards and regulations are slow in coming, the legal actions regarding IAQ and buildings have been accelerating. The number of claims that are currently in the court systems has increased since 1986 and the amount of those claims has ranged upwards of $10 million. In 1995, there was a $25.8 million judgment for the Polk County Courthouse, which had an incidence of building-related illness. Because of sick building syndrome, building-related illness, and other IAQ conditions that have developed in buildings, there are several other multi-million dollar law suits currently in process.

These suits are currently being settled for amounts ranging from several hundred thousand dollars to more than $1 million. The extent and award of these cases is also being facilitated by changes in how the awards are being made. Under some suits, the HVAC system has been classified as a product to allow the litigators to use product liability laws. There are additional legal costs that affect hospital administration that are not included in these costs: administrative support and investigation, examination and testing for problem determination, and HVAC system impact.

IAQ lawsuits against corporations are, for the most part, covered by either workers' compensation insurance (see Chapter 5) or product liability policies. Cases such as *Buckley v Kruger-Benson-Zeimeer (1987)* for $622,500,[20] *Shadduck v Douglas, Emmett and Co (1989)* for an undisclosed amount,[21] *Call v Prudential Insurance Co of America (1990)* for an undisclosed amount,[22] or *Bloomquist v Wapello County, Iowa (1990)* for $1 million[23] are becoming commonplace in building welfare suits.

A new trend in employee liability is criminal prosecution; for example, in an Illinois case officers of a nonmedical manufacturing company were indicted for homicide after a worker died from job-related poisoning. Clearly, it is in the hospital's best interest to consider the potential risk and costs of litigation in the evaluation of the IAQ associated from the lack of maintenance of the HVAC system and decreases in ventilation rates. The recommended steps[24] that health care organizations should take to limit their potential liability from indoor air pollutants are consistent with the IAQ program discussed in Chapter 6.

Workers' Compensation, Medical Bills, Health Care and Insurance Costs. Hospital administrators are privately expressing concerns, with good reason, that their hospitals may not provide the safe and healthy environment they once believed. NIOSH has published guidelines for evaluating occupational health and safety programs in hospitals; 8% of the 5,000 hospitals reviewed in a recent study are in compliance.[25] Many of their employees may be at risk due to environmental hazards such as exposures to toxic substances and poor IAQ. Exposures to potentially harmful contaminants can range from exposure to waste anesthetic gases to the "smoke" from laser surgery (see

Chapter 3), to toxic substances such as pentamidine and ribivarin when these substances are delivered in an aerosolized form.[26] In 1992, employers and workers' compensation insurers paid more than \$62 billion in worker-lost-time benefits and medical bills, according to the National Council on Compensation Insurance.[27] The most important aspect of keeping workers' compensation claims down is to prevent incidents before they occur.

Patient Health. IAQ problems in hospitals affect the hospital staff and can place a double burden on patients. They will be exposed to the same environment that the staff is, and any stress produced by compounds in the air may affect recovery time. Most office IAQ cost impacts are calculated against costs of employing workers. However, the "productivity" of the patient is focused on getting well, which means that if there is an IAQ problem in the hospital, the patient recovery time may increase.

Property/Equipment Damage and Maintenance. Damage to materials and equipment in hospitals by contaminants has only been minimally examined. The impact can range from ozone damage to rubber and organic compounds to the abrasive impact of particles on meshing surfaces. To date, the majority of research has been conducted in the HVAC system and points to the need to protect equipment, the impact of oxidants on materials, and the fungal defacement of finishes.

Electronic, examination, and laboratory equipment are also affected by IAQ. This damage will present itself in the reduced lifetime and increased maintenance of components, the oxidation of circuit boards, and the increased cleaning of laboratory, medical, and diagnostic equipment. Some of the effects on materials have been summarized by Heinan et al[28] including the tarnishing and corrosion of metals; the discoloration and soiling of paints and wall coatings; the reduced tensile strength, soiling, fading, and color change of textiles; microblemishes of photographic material; and the cracking and changes in surface appearance of paper, rubber, and ceramics.

The cost of more frequent washing and decreased lifetime of hospital clothing and bedding, and the increased effort added to normal building maintenance and cleaning, are elements not yet addressed in studies. The cost to provide increased cleanliness for indoor surfaces may result in about $100/maintenance person per year. If the life of garments and bedding decreases by only 0.1%, it can easily add to the total cost of the building operation.

In short, ensuring high-quality IAQ makes excellent economic sense in the health care environment. Health care organizations that choose to invest in maintaining good IAQ will find their dollars well spent.

References

1. Hymore C, Odom R: Economic effects of poor IAQ. *EPA Journal,* 28–29, Oct-Dec 1993.

2. Abdou O, Lorsch H: The impact of building indoor environment on occupant productivity Part 1: Recent studies, measures and costs. *ASHRAE Transactions* 100(2), 1994.

3. U.S. Environmental Protection Agency, Office of Research and Development: *Inside IAQ.* Research Triangle Park, NC.

4. Hallstron A: Designing an IAQ-ready air handler system. *Trane Applications Engineering Manual,* 5–7, Nov 1994.

5. Scary R: Looking into sick buildings. *Heating / Piping / Air Conditioning,* 31–34, Jul 1994.

6. Department of Labor, Occupational Safety and Health Administration: Indoor Air Quality; proposed rule. *Federal Register* 59(65): 15968–6309 Apr 5, 1994.

7. U.S. Bureau of the Census, Statistical Abstract of the United States: 1994 (114th edition.) Washington, DC, 1994.

8. Althof J, Monfre J: HVAC for today's hospitals. *Heating / Piping / Air Conditioning,* 47–50, Jul 1990.

9. Eto J, Meyer C: The HVAC costs of fresh air ventilation. *ASHRAE Journal* 30(9): 31 35, 1988.

10. Technical Solutions Guidebook for Performance Contracting, Internal Honeywell document, Second Edition, Jul 1996.

11. Steel T, Brown M: *ASHRAE Standard* 62–1989 energy, cost, and program implications. Bonneville Power Administration, 1990.

12. Woods J: Cost avoidance and productivity in owning and operating buildings. In Cone J, Hodgson M (eds): Problem Buildings: *Building Associated Illness and the Sick Building Syndrome* 753–70. Philadelphia, PA: Hanley and Belfus Inc, 1989.

13. Woods JE: Cost avoidance and productivity in owning and operating buildings. *Occupational Medicine: State of the Art Reviews* 4(4):753–70, 1992.

14. Abdou O, Lorsch H: The impact of building indoor environment on occupant productivity Part 3: effects of indoor air quality. *ASHRAE Transactions* 100(2), 1994.

15. Steele T and Brown M: 1990, ASHRAE Standard 62-1989, Energy, Cost and Program Implications, DOE/BP — 1657, Bonneville Power Administration.

16. Cyfracki L: Could upscale ventilation benefit building occupants and owners alike? *Proceedings of Indoor Air '90: 5th International Conference on Indoor Air Quality and Climate,* Vol. 5 135–41. Aurora, ON: Inglewood Printing Plus, 1990.

17. Holcomb LC and Pedelty JF: Comparison of employee upper respiratory absenteeism costs with costs associated with improved ventilation. *ASHRAE Transactions,* 100(2):914–21, 1992.

18. Collins JG: Health characteristics by occupation and industry: United States 1983 1985. *Vital Health and Statistics* 10(170), 1989.

19. Jones, MT: Liability concerns underscore the need to address IAQ issues. *Real Estate Weekly* 91(26): 406, Jan 25 1995.

20. *Buckley v Krueger-Benson-Ziemer.* Super. Ct. Santa Barbara, County, CA No. 1433393, 1987.

21. *Shadduck v Douglas, Emmett and Co.* Super. Ct. L.A. County, CA, No. WECI–36229, 1989.

22. *Call v Prudential Insurance Co of America.* Super. Ct. L.A. County, CA No. SWC 90913, 1990.

23. *Bloomquist v Wapello County, Iowa.* D.C. Iowa, Wapello County, No. CL 2785–0687, 1990.

24. Kirsch LS, Caldwalader, Wickersham, and Taft. Washington, DC.

25. Newman MA and Kahuba JB: Protect the health of our health care worker. (on-the-job health and safety issues), *Hospital & Health Services Administration* 36(4):537, 1991.

26. Charney W and Schirmer: *Essentials of Modern Hospital Safety.* Chelsa, MI: Lewis Publishers, Inc, 1990.

27. Davisson J: Cutting worker's compensation costs. *Occupational Hazards* 56(2):26–29, Feb 1994.

28. Heinen DC, Rundus RE, Kemme MR, Nemeth RJ, Kinzer GR, Quintan KA, Adeoye BF, Rives BL: *How Building Systems Affect Worker Wellness,* USACERL Technical Report, FF-94/17, 1994.

Chapter 9: IAQ Program Case Studies

Many health care facilities have implemented IAQ programs for a variety of reasons other than code or regulation requirements. Indeed, at the time that these programs were started, there were no requirements for IAQ programs.

The following three case studies have been provided by pioneering health care facilities that have had an IAQ program in place for several years. These case studies are somewhat different in their scope and intent but all summarize their overall experience with the IAQ program.

A. Torrance Memorial Medical Center (TMMC)

Bud Maxfield, Director of Facilities

Mr Maxfield has a degree in engineering, and more than 35 years of experience in facilities engineering and construction management. He has been with TMMC, Torrance, California, for 17 years, nine years as Construction Manager, and eight as Director of Facilities.

Torrance Memorial Medical Center has always worked to stay on the cutting edge of technology, and was pleased to be rewarded for these efforts by being named one of the Top 100 Hospitals in the nation in 1993, 1994, and 1995 by the consulting firm of Mercer, Inc. Our Facilities Engineering Department has always enjoyed a harmonious working relationship with the Infection Control and Employee Safety Departments, and it was at one of these meetings in 1993 that we discussed the IAQ issue.

We had all had some exposure to the issue from presentations made at the professional society meetings that we attended and from articles in the various journals. At that point, we had not received any IAQ-related complaints that we could not address, but the increase in frequency of these complaints made it clear to us that we needed to take more formal action to prevent the kind of surprises that no hospital wants.

After several in-house meetings and discussion with administration members, we contacted an IAQ consulting firm for assistance. The consultant guided our loose knit group into what has become our IAQ management team, and working together, we designed an IAQ program for our facility. We did not have to search for areas to use this new resource, as our facility seems to be in a constant state of renovation, expansion, remodeling, or some other form of construction. We were also undergoing a major mechanical system analysis, design, and construction upgrade from an energy conservation perspective and had several projects about to start throughout our facility. Our Construction and Engineering Departments sought input and guidance from our newly formed IAQ management team, thereby enabling the hospital to avoid many pitfalls that may have resulted in compromised IAQ and complaints. Those complaints that we did receive were quickly addressed.

We know that our IAQ program has been of great value to TMMC and is a sound business decision. A few examples will illustrate this:

1. One of the first issues to develop was in our Laboratory area. This department is at the end of the run for the supply air system and was in need of additional capacity. We periodically responded to comfort calls and a few complaints of odors, due in no small part to the capacity issue or so we thought. Our new IAQ management team undertook an assessment of the area, and discovered that an old transfer fan disassembled in the course of maintenance had never been hooked back up; a surprise to us. This discovery was very timely, since we had major system renovation planned for this area. Adding the project of hooking the fan back up to the planned upgrades just made good sense.

2. Medical records areas in all hospitals tend to be complaint generators, and ours is no exception. When our team investigated these complaints, they found that ductwork had been crushed over the years, restricting airflow to this area. Again, because of planned HVAC system modifications, the recommendations were appropriate and timely.

3. Some of our top administrators are located on the second floor of the main building. We had received complaints from one of these individu-

als, which we took very seriously in part because this person has a reputation of never complaining. Our IAQ team discovered several issues pertaining to original construction that needed to be addressed. For example, the building was originally planned to have sunshields over the windows. These were eliminated due to budget constraints. Therefore, the building temperature would follow the sun, affecting airflow throughout the HVAC system. As in other cases, we had planned construction in this area as well. If we had not had the input from our IAQ management team, these recommendations would not have been incorporated into the renovation, and we would still have had the problem.

4. Our MRI unit is attached to the main building but is served by a separate air conditioning unit. Complaints in this area pointed out deficiencies in the way some of the equipment was being used and a negative influence from construction of a new building adjacent to this area. This led us to undertake training of personnel in the use of the equipment, and to make some changes in the way our construction crew accessed the area.

5. A situation developed in our East Wing Critical Patient Care area that required immediate attention: the possibility of a nosocomial infection. Fortunately, this was not the case, but it was still investigated by our team. In performing their assessment, the team found that some water had leaked into the HVAC system, resulting in an increase in bacteria. Because of the original possibility of a nosocomial infection, the time between the initial complaint and commencement of remediation was extremely short—less than 48 hours. During the period of time that we received this complaint, we were also in the planning stages of making major changes to the air-conditioning system for this tower to accommodate past growth of the functional use of the building. Like many facilities in this situation, we had grown beyond the capacity of our HVAC system's ability to serve us. Our consulting mechanical engineer and architect were in the process of surveying our system needs and had prepared a preliminary cost estimate for updating the

system. Because of the findings, they made duct cleaning a part of the plan.

All of these incidents were valuable to our IAQ program in the following ways:

- Maintaining and improving the quality of our facility for our patients and our employees;

- Maintaining and improving our relationships with the medical staff and other function centers in the hospital;

- Improving morale and productivity; and

- Avoiding costs from employee absenteeism, workers' compensation claims, and unnecessary construction.

Although all of these are true benefits, it is difficult to quantify them with actual dollar savings. The exception to this statement is example 5, which, much to our surprise and delight, has actually given us hard data for cash savings.

One of the recommendations from the IAQ management team, based on testing within the air-conditioning system and on the patient care floors, was to clean the HVAC system. This work was begun immediately. It was coordinated with the other survey work being done by the consulting mechanical engineer, and the selected mechanical contractor, a necessity because the areas served by the HVAC system were occupied 24 hours a day and could not be off line for more than six hours during the evening. The mechanical engineer was preparing a contingency plan to provide temporary air-conditioning, if it was required.

As part of their scope of work, the mechanical engineer hired a certified air balance agency to take performance readings of the HVAC system. Using these readings and the calculations that their office had prepared, they were able to determine the amount of additional capacity that would be required to meet our needs. This information was given to our IAQ management team for analysis. Our IAQ consultant reviewed this data, along with the plans and specifications for the original construction and design capacities.

After the IAQ consultant completed his review and analysis, he determined that cleaning the HVAC system would restore lost capacity and that very little if any additional capacity would be required. This analysis was given to our consulting engineers and architects, and meetings were set to discuss this issue. The two consultants had a meeting of the minds on this, and it was agreed that the mechanical consultant would have a certified air balance agency repeat the system survey after the cleaning project was completed.

The results of the air flow readings are given in Table 9–1, below. (It must be pointed out that our facility has never met the original design specifications for air flow.) The first air balance report made after the building came on line showed supply and return air deficiencies.

Measurement	Supply Air	Return Air	Exhaust Air
Original Design	64,340	42,660	13,840
Before Cleaning	43,854	21,413	16,094
Design Deviation	-20,486	-21,247	+2,254
After Cleaning	59,195	30,115	17,568
Design Deviation	-5,145	-12,545	+3,728
Net Gain	15,341	8,702	1,474

Table 9–1: Air Measurement Readings at TMMC

The air balance data sheets also showed changes in the static pressure across the HVAC unit, and the full load amperage measured at the 50 horsepower fan motor. These data are summarized in Table 9–2, below. The mechanical consultant has reviewed the complete balance report, and has determined that we have enough capacity in the existing system

Measurement	Static Pressure, Inches HG	Full Load Amperage Average of 3 Legs
Before Cleaning	3.85	53
After Cleaning	3.07	50
Decrease	0.78	3

Table 9–2: HVAC System Measurement at TMMC

to meet our needs. All that will be required is to rebalance the system to the new air flow values calculated by the engineer.

After some minor ductwork is revised and all of the balancing work is completed, our engineer estimates that we will save $300,000 from avoided equipment purchase and installation costs.

In examining these data, it is clear that if we had not been in need of additional capacity, we could have affected energy savings by simply adjusting the fan back to the precleaning air flow values. This would also have allowed us to achieve additional savings by resetting the chilled water temperature. Clearly, these are benefits that we never imagined we would be getting from our IAQ program.

To summarize the experience at TMMC, we feel that our program was a sound business decision. Rather than a cost of doing business, we view it as a benefit because we are staying on the cutting edge.

B. Scottsdale Memorial Hospitals

Kenneth E. Lewis, MPA, Safety Officer

Mr Lewis directs the safety and industrial hygiene programs at the hospital in Scottsdale, Arizona. In addition, he has nearly 20 years of experience in hazardous materials management and oversees the corporatewide Hazardous Materials and Waste Management Program.

Most indoor air quality programs at health care facilities are created out of need. They begin when administrators become aware of employee concerns about the air quality in their departments. The concerns range from strange odors to reports of feeling sleepy in the afternoons to questions about possible health effects.

On two separate occasions we hired an outside consulting firm to test the air in two of our buildings. This proved to be a costly and reactive approach. To be more proactive in our approach, we decided to perform most of the IAQ testing in-house. To enable us to perform these functions, we joined a regional organization dedicated to IAQ. We were able to learn more about IAQ and develop a plan to purchase equipment to perform testing. Training was also available through various organizations, ranging from OSHA to university-level instruction.

The training allowed us to make educated decisions concerning the purchase of test equipment. Because we have equipment available to us all the time, we are able to test areas whenever a problem occurs. Another advantage we found is that timely response to suspected problems eases our staff's concern.

As most people in the IAQ business realize, there is often a high psychological aspect to IAQ problems. When one person feels that there is an IAQ problem, others may soon feel the same way. The ability to test for basic IAQ parameters allows us to know rather quickly if there is a problem. Employees appreciate having a timely response to their concerns and many times feel better just because someone is looking into the situation.

Most of the IAQ problems we have found relate to the HVAC system. The majority were solved by simply bringing more fresh outside air into the building. This lowered the carbon dioxide level in the building and the employees, most of the time, could sense improvement in the quality of the air.

Another area of concern related to IAQ during construction projects, ranging from asbestos removal and containment to the potential of liberating bacteria during a project. Our health care system has been in a growth period and numerous construction projects have taken place in the hospitals.

Our safety committee developed a special policy to specifically address construction issues. Members from Construction, Facilities Engineering, Infection Control, and Safety met to look at possible problems and to develop a workable policy to protect our patients, visitors, and staff during construction projects. The policy was designed to meet long- and short-term projects.

We reviewed policies of other health care providers and an article published by the American Industrial Hygiene Association (AIHA) dealing with air sampling during construction. The result was the following workable policy that was approved by the safety committee.

Safety Committee Policy on Control of Construction Airborne Contaminants

The purpose of this policy is twofold: (1) to provide direction for safe patient transport through or around a construction zone, and (2) to identify precautions for controlling construction-related airborne contaminants (including dust, airborne fungi, vapors, and odors).

■ *Patients susceptible to airborne contaminants (mold, bacteria):*
   Patients confined to our health care facilities are considered to be susceptible to infection. We take numerous steps to reduce the likelihood of patients being exposed to infectious agents.

   • This policy will set forth methods that shall be taken during construction projects to minimize exposure to airborne contaminants.

■ *Types of projects that require barrier precautions:*

- Demolition of wall board, plaster, ceramic tiles, or ceiling tile;

- Demolition of flooring;

- Removal of windows/doors;

- Removal of casework; and

- Any construction activity that will cause dusts to be liberated.

■ *Types of appropriate barriers:*

- Closed doors, with tape applied over the frames and doors, will be acceptable for a project which can be contained within a single room.

- Long-term barrier (projects greater than one week)—Demolition of walls, floors or ceilings will require a barrier to contain the metal or wood studs and plywood or wall board. It should extend from the floor to the ceiling, including the area above ceiling tiles. Seams should be sealed with tape. If it is impossible to seal above the ceiling space with plywood or wall board, a plastic barrier should be used.

- Short term barrier (projects less than one week)—A barrier constructed with plastic and studs will be acceptable for projects of this duration. If the project involves above-ceiling work, the barrier should contain the above-ceiling space. This plastic barrier must be air-tight at all times. Any penetration through plastic must be repaired immediately with tape.

- Dust control within the construction area—Damp mopping should be used for dust control. Refuse being disposed of should be transported internally using covered containers. Tacky mats should be used at the entrance to the construction area.

■ *Air handling considerations during construction:*

- External construction;

- Exhaust from equipment should be directed away from all building air intakes. Gas-powered equipment placement should be reviewed by the Project Manager.

- Dirt excavation or building demolition should not occur unless building air intake filters are properly installed and operational. Maximum recirculation of building air during external construction will minimize construction dust penetration.

- Construction aerosols (dust) should be reduced by sealing windows adjacent to construction zones. Cracks around windows and door frames should be caulked or weather stripped to reduce infiltration.

- External construction debris should not be transported through hospital buildings.

- Internal construction projects should be barrier-protected. The type of barrier required should be evaluated by the Project Manager.

- Air should flow from the clean area to the dirty area (negative pressure in the construction area). Control should be implemented utilizing the exhaust building ventilation or window exhaust. To assure proper clean-to-dirty air flow, smoke stick evaluations should be conducted (prior to the start of the projects) by the project manager.

- Windows within construction zones should not be opened during construction, unless used for exhaust ventilation. After hours, all windows should be closed.

▪ *General considerations during construction:*

- Debris transported from internal demolition should be routed to avoid areas housing patients and visitors where possible. Elevator usage and type of transport container should be designated by the hospital facilities office.

- No food is to be allowed in the construction areas.

- Active demolition (wall removal, utility work, etc.) requires evaluation of occupants above, below and on either side, for their susceptibility to airborne contaminants.

- All through-the-wall connections, such as utilities and ducts, should be sealed appropriately and reviewed by the project manager.

- When demolishing dust-producing materials (for example, plaster, wall board, ceiling tiles), the areas should be misted with water to minimize dust.

■ **Environmental services considerations during construction:**

- The housekeeping supervisor should be notified by the facilities and/or construction office prior to the start of an internal construction project, that arrangements for damp mopping of floors around the area can be scheduled. This mopping should be done frequently enough to prevent tracking of dust to adjacent areas.

- Prior to the final acceptance, Housekeeping should be notified by the Hospital Facilities/Construction office to begin final cleaning of a new construction zone.

■ **Direct patient/staff responsibilities for patient transport:**

- Staff need to be alerted to the location of construction so that, when transporting patients, the construction zones and routes are avoided when possible.

- Patients with decreased white counts should wear face masks when being transported near construction zones.

IAQ has shown itself to be an ever changing program. The OSHA tuberculosis standards now must be taken into consideration when reviewing IAQ concerns; the propsed OSHA regulations on IAQ must also be considered.

Planning and involvement with construction, facilities engineering, and your HVAC personnel are vital in order to design, build, and maintain the proper IAQ. Even though we are satisfied with our IAQ program, we know that we do not have all the answers. We still utilize an IAQ consultant when issues exceed our technical expertise.

We have found, however, that developing a program and timely response to IAQ complaints has paid tremendous rewards in improved morale and our ability to provide a workplace conducive to productivity and safety.

C. Kaiser Foundation Hospital

Fred Brunsmann, Chief Engineer

*Mr Brunsmann is a plant engineer at Kaiser Foundation
Hospital, Panorama City, California. He has more than 38
years of experience in facilities maintenance and operation. He
has been with Kaiser for over 25 years.*

At Kaiser Permanente, we have always prided ourselves on taking the proactive approach to facilities management, and it has continually paid dividends for us. In the past, our facility has had some complaints or questions about the building environment, to which we readily responded—at least to the best of our knowledge and abilities. If that wasn't sufficient, we assumed that the complainant had a personal problem. Then IAQ issues came to the forefront, and in the spring of 1991, our approach to handling employee IAQ complaints changed.

One of our medical office and clinic buildings was undergoing some remodeling and renovation. During this period, we found temporary quarters for some of our home health care nurses in a part of the building that was not under construction. This room had been an examination room with an adjacent restroom. Shortly after the nurses moved into this space, we began to get complaints of odor and discomfort. The complaints were severe enough that the medical doctor ordered them home. Our engineering staff responded quickly and investigated the area, but to no avail. During this investigation, the complaint moved to a union grievance and then to a workers' compensation claim.

Our next step was to enlist the aid of a consulting firm to perform an IAQ survey and investigation. Air testing revealed hydrogen sulfide and miscellaneous volatile organic compounds in the space, along with elevated levels of carbon dioxide. Within a matter of a few hours, the path of the pollutants and the cause had been found, and recommendations were given to our engineering staff. We were able to make the corrections in-house in a matter of two days.

This experience was an eye-opener and a wake-up call. If we were to maintain our proactive philosophy, we needed to broaden our scope of vision.

How the IAQ Program Developed

We met with the principals of the consulting IAQ firm that had solved our problem to see what we could do to avoid similar episodes in the future. This was the beginning of a very beneficial learning experience for Kaiser Permanente. Given all of the functions and systems in a full hospital, medical office buildings, and clinic buildings, we had visions of all sorts of horrors besetting us. Working with the consultant, we developed a program that initially addressed only the main hospital, but has since evolved to include all buildings attached to our facility. This program included:

■ Complete survey and inspection of all air handling systems. This included a visual and photographic inspection of the interior of the units and duct systems.

■ Airborne background testing of facility. Tests were made at areas of the building corresponding to each air handling unit.

■ Complete report with recommendations.

After the report recommendations were carried out, the second phase of the program began. In this phase, the consultant began to periodically choose selected areas of each building for testing and to conduct a survey of the air handling unit serving that area. We never knew which areas would be selected, or when the consultant would be testing in these buildings.

As mentioned previously, our program has changed very little since its inception, other than to include additional buildings. However, one enhancement that bears mention is the addition of a Rapid Response Service. We occasionally have an incident, as all hospitals do, in which we need immediate help: a glutaraldehyde issue, "mystery odor," employee complaint, and so on. While we have become more knowledgeable and sophisticated regarding IAQ in the past six years, we still have the occasional need of an outside expert. With the added Rapid Response Service, we have had our consultant on site within two hours of the call, which has been very helpful.

We have also used this program as a method of quantifying the need for certain maintenance procedures such as internal cleaning of the air-conditioning system. Kaiser had a history of doing this type of work prior to the program, but now it is handled on a more routine and programmed basis.

When we compare where we were in 1991 with where we are today, we can see the improvements in our program and our handling of it. As a side note, our program exceeds the suggested requirements of the proposed OSHA regulations, and those shown in the ASHRAE 62-1989R IAQ Standard.

The IAQ Program Experience: Integration into our System

We had some concerns initially over how this program would be managed and how much additional work it would involve for our already overworked engineering staff. We also had budgetary concerns, because we had no budget for a program of this type. Interestingly enough, the funding was far easier than we anticipated. We discussed the program with the Employee Safety/Risk Management Committee and the Infection Control Committee and found strong allies in both of these groups. We have always had the highest regard for the safety of our employees, and the quality of service provided to our members. When the program was discussed from this point of reference, it became a logical business decision. In fact, the united front that we presented to management was what sold the program.

Integration of the program into our existing structure was almost effortless, once we had an understanding of the program and the flow of information. Within a few months, we were almost unaware of the program in terms of our daily routine. Occasionally, our consultant will call us about a particular issue that needs our immediate attention, but that is a welcome break in our routine.

When we look at the inspection and record keeping aspects of the proposed OSHA regulations, and compare them with what has become our normal routine, we see little cause for concern. The fact is, all hospitals have a higher level of routine maintenance and equipment inspection than does the typical commercial enterprise.

IAQ Program Benefits

As we prepare our annual budget, we do tend to lose sight of the IAQ program benefits. Preparing this case study has been beneficial, as it has allowed us to review where we were, how far we have come, and where we are going.

As one might suspect, quantifying the results of our program with a dollar amount of benefit is very difficult. Most of the benefit comes from avoided costs, rather than actual cost savings. However, one area where we have actual data comes from our Continuing Care Department. This unit is in leased quarters away from our main campus. We had numerous complaints in this area related to IAQ and general malaise when the employees were in this building for any length of time. The administrator of that department logs all employee absences and the reason for those absences. Prior to the beginning of our IAQ program, the Continuing Care Department showed an average of 168 hours of employee absences per month for lung and respiratory complaints. After the IAQ program was put in place, these absences decreased to seven or eight per month. In the five years since the inception of this program, we have had two IAQ-related complaints from this department, which were very minor and quickly dealt with.

We have also had instances in other leased quarter facilities where employee concerns were quickly addressed by having the consultant's Rapid Response Service meet with them directly and investigate the complaints. The most prominent of these involved temporary quarters in a mini-mall type of structure. (The reason for this temporary move was because earthquake damage required us to vacate a brand-new two-story facility.) The two instances in this leased facility were:

1. *Strong, noxious odors in the optical department.* The investigation showed that rodents had entered the ceiling plenums, which were com-

mon to other tenants in this building, the most notable being a restaurant. The owner of the building had been notified of this problem by the restaurant owner, who then placed rat poison in the ceiling plenum. The consultant's team identified the problem, and embarked upon a mitigation procedure within two hours of being notified by our engineering department.

2. *The infiltration of volatile organic compounds into our Urgent Care Department from neighboring tenants.* This forced the evacuation of our staff. This problem was traced to a dry-cleaning establishment located at the end of the building that was operating without benefit of proper ventilation and exhaust. Again, this problem was identified and rectified within two hours of notification. In this case, the consultant worked with our leasing and legal department to effect a correction with the owner that we feel saved us additional legal costs.

There have been other questions that periodically surface regarding exhaust systems in sterilization rooms and glutaraldehyde-related issues. These have been expeditiously dealt with by our IAQ team as the result of the education we have received from our consultant and the IAQ program.

Other instances in which we feel the IAQ program paid for itself are in the area of employee complaints, union grievances, workers' compensation claims, and OSHA-related issues. Our IAQ-related complaints have dropped by more than 90%. We had one OSHA complaint after the program began, which was dispelled when the OSHA inspector reviewed our program and met with our consultant.

In summary, we feel that our IAQ program was a good business decision and is consistent with our corporate goal of providing high-quality service to our members and to our employees.

Part III: Sample Planning and Monitoring Forms

This section contains several sets of forms that have been used by health care facilities to assist in their IAQ assessments. These forms are presented for readers' use as a guide in performing their assessments and setting up their own IAQ programs.

Appendix A. Indoor Air Quality Management for Hospitals Checklist

Andrew J. Streifel, MPH, REHS

Clinical Issues for IAQ

- Environmental microorganisms
    - ❏ Monitoring for sources of patient related infections
        - — *Aspergillus* or other airborne opportunistic fungi
        - — *Legionella*
    - ❏ Infection recognition
        - — Response by personnel to include assurance of ventilation or water quality parameters (CDC)
    - ❏ Protective measures
        - — Ventilation management, protected environments
        - — Operational procedures
    - ❏ Environmental inspections
        - — Cleanliness, water damage control, windows, construction/renovation/maintenance protocols
        - — Critical areas: operating rooms, central sterile processing, disinfection/sterilization, patient rooms, pharmacy, kitchens
- Patient derived airborne infection management
    - ❏ Recognition of patients with infection
    - ❏ Engineering controls for airborne infection isolation
    - ❏ Monitoring for ventilation control assurance

Volatile Organic Chemical Management

- Location identification for chemical control
    - ❏ Glutaraldehyde fumes
    - ❏ Ethylene oxide gas
    - ❏ Waste anesthetic gas

❏ Laboratory chemicals

- Ventilation management for hazardous chemical control

    ❏ Enhanced air exchange for usage area

    ❏ Exhaust ventilation

    ❏ Greater exhaust versus supply air volume

<u>Emergency Response</u>

- Spill management

    ❏ Chemical spill response team

    ❏ Hazmat training

- Water damage response

    ❏ Immediate drying or removal for mold growth prevention

- Clearance determination for safe reoccupancy

    ❏ Air quality assessment for hazardous material

Protective Environments Checklist

_____ Pressure measurement (monitoring)

_____ Air exchanges per hour

_____ Self-closing doors

_____ HEPA filtered particle counts

_____ HEPA filtered airborne fungi sample

_____ Room set up for clean to dirty airflow

   (patient considered "clean")

Airborne Infection Isolation Rooms Checklist

_____ Pressure measurement (monitoring)

_____ Air exchanges per hour

_____ Self-closing doors

_____ Room set up for clean to dirty airflow

   (patient considered "dirty")

Construction Management Planning Checklist

_____ Risk area

_____ Ventilation outages

_____ Demolition control

_____ Volatile chemicals

IAQ Complaint Management Checklist

_____ Phone number for response (infection control, safety, or maintenance)

_____ Category of investigation

_____ Employee health involvement

_____ Security utilization for vehicle control

_____ Investigation and response

Appendix B. Fungal Growth Checklist

Philip R. Morey, PhD, CIH

Moisture Issues in HVAC Systems

(1) Yes ____ No ____ Standing water occurs in drain pans. If yes, redesign pan for complete drainage. (Chapter 2, ref. 17)

(2) Yes ____ No ____ Rust and corrosion occur on airstream surfaces in air handling unit plenums. Rust and corrosion indicate chronic moisture problems and niches for microbial growth.

(3) Yes ____ No ____ Chronically damp porous materials occur on airstream surfaces. Wet or chronically damp filters, insulation, and other materials susceptible to biodeterioration should be replaced. Airstream surfaces should be smooth, easily cleanable (or replaceable), and resistant to biodeterioration.

(4) Yes ____ No ____ Water droplet carryover from cooling coils, water spray systems, and humidifiers occurs. If yes, redesign or add demisters for zero water droplet carryover.

(5) Yes ____ No ____ Water droplets from cooling towers or from fogs or mists are entrained in HVAC outdoor air inlets. Relocate outdoor air inlets to minimize entrainment from cooling towers. Horizontal and vertical separation distances as well as air flow patterns around the building and prevailing wind direction are important considerations in outdoor air inlet relocation. In climatic zones with chronic fog, installation of preheat coils may be an option to decrease moisture entrainment.

(6) Yes ____ No ____ Stagnant water occurs in humidifier reservoirs. The presence of *Pseudomonas* spp., *Flavobacterium* spp., and yeasts in stagnant water is very likely. Upgrade the preventive maintenance of the humidifier (daily dumping of sump water and disinfection of wet surfaces) or replace unit with a system that emits steam or water vapor (no water droplets).

(7) Yes ____ No ____ Airstream surfaces near humidifiers are wet. Wet surfaces indicate a leak from the unit or that moisture is not absorbed by the ventilation airstream. Redesign unit so that leaks do not occur and that all emissions are totally absorbed by the airstream.

Condensation in Occupied Spaces

(1) Yes ____ No ____ In hot, humid climates or seasons, condensation or dampness occurs on the inner surface of envelope walls. If yes, avoid conditions such as overcooling, negative pressurization, and use of wall coverings impermeable to moisture. (Chapter 2, ref 2)

(2) Yes ____ No ____ In hot humid climates or seasons, condensation or dampness occurs on chilled water pipes or on surfaces (for example, supply air vent louvers) chilled by conditioned air. Add insulation to surfaces of chilled water pipes, redirect supply air away from surfaces being overcooled, and verify that HVAC equipment is capable of removing latent (moisture) loads from indoor air.

(3) Yes ____ No ____ Condensation or dampness occurs on envelope walls in cold climates or seasons. If yes, eliminate thermal bridges, improve insulation in walls, and provide ventilation (with outdoor air; in cold seasons outdoor air generally has a low moisture content) to perimeter zones.

(4) Yes ____ No ____   Condensation or dampness occurs on envelope walls in humidified zones in cold climates or seasons. For very cold climates a room RH of 35% or more may result in dampness or condensation on or in envelope walls. Determine if thermal gradient across the envelope wall will result in surface dampness or condensation. Reduce humidification of room air or improve insulation in the wall. (Chapter 2, ref. 2)

Floods and Moisture in Occupied Spaces

(1) Yes ____ No ____   A written plan is available for drying interior finishes and construction material within 24 hours after any flood including those caused by water pipe or fire sprinkler rupture. Drying is considered adequate if surface relative humidity of construction and finishing material is less than 65% (Chapter 2, ref 15). Dehumidification equipment must be on hand or available to provide for adequate moisture removal from porous materials.

(2) Yes ____ No ____   Room RH consistently exceeds 60%. Air in rooms where RH consistently exceeds 60% will be perceived as uncomfortable by occupants.

(3) Yes ____ No ____   Surface RH of porous finishing materials (carpet, wallboard, insulation) consistently exceed 65%. Materials susceptible to the biodeterioration should be avoided in these areas. Mold growth will occur when surface RH exceeds 80% for a few days. (Chapter 2, ref 16, 17)

HVAC System Preventive Maintenance

(1) Yes ____ No ____   Outdoor air inlets are protected from construction

ducts, vehicular emissions, and other contaminants.

(2) Yes ____ No ____   A plan exists that minimizes dissemination of dust and soot from airstream surfaces in air handling units or ducts when preventive maintenance or cleaning occurs. Point of discharge filters on air supply vents must be considered to protect immunocompromised patients from HVAC dusts.

(3) Yes ____ No ____   Inspection shows that dust and soot have accumulated on airstream surfaces in air handling units and ducts. The presence of dust and soot is an indication of deficiency in HVAC filtration and cleaning. It should be assumed that thermotolerant *Aspergillus* spp. are present in dust and soot deposits.

(4) Yes ____ No ____   Measurements made by direct reading particle counters show that concentrations of respirable particles are low in air downstream of filters. Air may bypass (flow-around) filters or filters may be absent (for example, not installed) in filter frames (see Table 2-5). A false sense of security may be assumed just because highly efficient filters are supposed to be present.

Renovation and Housekeeping

(1) Yes ____ No ____   A plan is in place and enforced to prevent dust generated by renovations from entering all patient and general use areas. If no, see recommendations in Chapter 2, references 2, 9, 15, and 19.

(2) Yes ____ No ____   Protocols controlling renovation activities incorporate stringent dust control measures. The use of rigid, impervious, slab to slab critical barriers and

the negative pressurization of all work areas is required. Discharge of air from work areas must be through HEPA filters and all work areas must be negatively pressurized at all times to interior patient and non-patient areas.

(3) Yes ____ No ____ Potted plants and other natural materials whose surfaces may be contaminated by culturable fungi are excluded from areas housing immunosuppressed patients.

(4) Yes ____ No ____ Cleaning of floors is performed in a manner that does not aerosolize dusts or leave surfaces chronically damp.

Dust from Outdoor Sources

(1) Yes ____ No ____ A plan exists and is enforced to suppress dust generated during soil excavation, construction, and demolition in locations in the vicinity of the medical facility.

(2) Yes ____ No ____ Administrative controls are utilized to prevent tracking of outdoor construction dusts into the medical facility by visitors, patients, and staff. Special consideration should be given to potential exposures to construction dust by immunosuppressed patients entering or leaving the facility.

(3) Yes ____ No ____ HVAC outdoor air inlets are protected from entrainment of dust during times when soil excavation, building demolition, and construction occurs. Protection may involve additional filtration, changing filters, and turning off air handling units when extremely high dust levels are expected. (Chapter 2, ref 9)

Presence of Visible Fungal Contaminants

(1) Yes ____ No ____ Visible fungi and/or bird droppings are present. If yes, assume that *Aspergillus fumigatus* and/or

Cryptococcus neoformans are present. Contaminants must be removed and surfaces disinfected using stringent containment procedures (Chapter 2, ref 9, 15, 19).

(2) Yes ____ No ____   Dust from visibly moldy areas and/or from bird droppings has been disseminated into the HVAC system or interior areas. Fine dust must be removed from affected HVAC or interior surfaces using HEPA vacuums, damp wiping, and dust containment procedures (critical barriers and negative pressurization). The objective of clean-up activities is not to sterilize building surfaces but rather to reduce levels of culturable fungi to background concentrations typically present in well maintained facilities.

Biofilm and Legionella

(1) Yes ____ No ____   Biofilm is present in drain pans and humidifiers and on other wet surfaces. If yes, physical removal and disinfection (HVAC component turned off) is required. Upgrading of preventive maintenance is also required so that biofilm does not reoccur.

(2) Yes ____ No ____   The cooling tower water system is treated to control microbial growth and a maintenance record of the water treatment program is available for inspection. If no, see appropriate publications (Chapter 2, ref 5, 6, 7) for guidance on maintenance of cooling towers.

(3) Yes ____ No ____   Cooling tower water systems are periodically monitored for culturable *Legionellae*. If no, consider testing as a means of verifying that the preventive maintenance program is controlling amplification of *Legionellae*.

(4) Yes ____ No ____   A protocol is in place for emergency decontamination of cooling tower water systems if an outbreak of Legionnaires' Disease occurs. If not, consult appropriate publications (Chapter 2, ref 5, 6, 7) for guidance.

(5) Yes ____ No ____   A written preventative maintenance and inspection program is followed so as to minimize the risk of *Legionella* growth in portable water systems including calofiers, storage tanks, plumbing lines, and outlets such as shower heads. If no, see Chapter 2 references 5, 6, and 7 for guidance.

(6) Yes ____ No ____   Portable water systems serving immunocompromised patient areas are periodically monitored for culturable *Legionellae* to verify the effectiveness of the preventative maintenance program.

Appendix C. Facility Mechanical System Inspection and Service Checklist and Log Sheet

NIOSH/EPA Building Air Quality Handbook

The following three pages are variations of the forms found in Chapter 6, reference 4. They have been modified from the original for use in other IAQ programs, and are offered here as a guide.

IAQ MAINTENANCE CHECKLIST FOR HVAC SYSTEM AND COMPONENTS

BUILDING: ____

A/C SYSTEM:_____________________________ DATE:_________________

TECHNICIAN: ____

SIGNATURE:____

COMPONENT OR ELEMENT	ACTION OR COMMENT
Unit Casing: All sections airtight and clean inside, all access panels secured.	
Outside Air Section: Check for any obstructions at louver, and proper damper operation.	
Mixing Plenum: Proper damper operation.	
Filter Section: All filters in place and snugly in frames, no bypass, no signs of mold on back side.	
Cooling Coil Section: Coil finned surface clean and free of debris, no indication of carryover, pan clean and draining, bio-cide tablet in place.	
Supply Fan Section: Fan clean and running smoothly, no belt slippage.	
Heating Section: Area and heat transfer surfaces clean and fee of debris.	
Controls: Operating in accordance with sequence of operation, no malfunctions.	

(Continued on next page)

IAQ MAINTENANCE CHECK LIST FOR HVAC SYSTEM AND COMPONENTS *(Continued)*

BUILDING: ____

A/C SYSTEM:_____________________________ DATE:_______________

TECHNICIAN: ____

SIGNATURE:____

COMPONENT OR ELEMENT	ACTION OR COMMENT
Supply Ductwork: Signs of visible contamination; dust, mold, other debris. Signs of leakage or water in duct.	
Terminal Units: Re-heat and vav: signs of contamination inside, controls functioning signs of water, minimum set-point holding.	
Thermostats: Functioning and calibrated properly.	
Return Air Plenums: Ceiling tiles in place, no signs of contamination above tiles.	
Exhaust Fans: Operating properly, controls functioning.	
HVAC System Standard PM: Refer to vendor's log for any corrections needed or done.	
Cooling Tower Standard PM: Refer to vendor's log for water treatment and any corrections needed or done.	

Source: NIOSH / EPA BUILDING AIR QUALITY HANDBOOK, Chapter 6, reference 4. Adapted by permission.

HVAC SYSTEMS LOG SUMMARY*

DATE	BLDG	HVAC NUMBER	MECH AREA	MAJOR EQUIP	FAN UNIT	TEMP CONT	AIR DIST	TECHNICIAN

* NO DISCREPANCIES FOUND = 1

CORRECTIVE ACTION TAKEN = 2 (REFER TO FORM C-1 FOR DETAILS)

Source: NIOSH/EPA BUILDING AIR QUALITY HANDBOOK, Chapter 6, reference 4. Adapted by permission.

Appendix D: IAQ Complaint, Interview, Diary, and Log Sheets

NIOSH/EPA Building Air Quality Handbook

The following four pages are variations of the forms found in Chapter 6 reference 4. They have been modified from the original for use in other IAQ programs, and are offered here as a guide.

INDOOR AIR QUALITY COMPLAINT FORM

DATE: ____

Please fill in this form and return it to:

____IAQ Program Designated Person

NAME:____

PHONE: ____

BLDG: ___________ DEPT: _________________ AREA/ROOM: __________

BEST TIME FOR US TO CONTACT YOU:_______________________

Please describe the nature of your complaint, when it began, and when you notice it most.

____

____

____

____

____

____

____

____

____

____

____

____

FOR OFFICE USE ONLY

FILE NUMBER: ______ RECEIVED BY: ______ DATE: ______

Source: NIOSH/EPA BUILDING AIR QUALITY HANDBOOK, Chapter 6, reference 4. Adapted by permission.

INDOOR AIR QUALITY INTERVIEW FORM

DATE: _______________ INTERVIEWER: ___________________________________

Interviewer: Please return form to IAQ Program Designated Person

EMPLOYEE: _______________________ PHONE: ______________________

BLDG: ___________ DEPT: _________________ AREA/ROOM: ___________

DATE IAQ COMPLAINT FORM FILED: ___________________________________

Description of complaints

Upper respiratory?: _______ Symptoms: ___________________________________

___ Eye irritation   ___ Skin irritation   ___ Headache   ___ Fatigue

___ Odor   ___ Inability to concentrate   ___ Nausea

___ Room temperature or humidity   ___ Sense of stuffiness in room

Other; please describe: ___________________________________

____

Do you have any health issues that make you susceptible to IAQ issues?
___ contact lenses   ___ allergies   ___ pregnancy   ___ cardiovascular disease
___ chronic respiratory disease   ___ medical treatments (chemotherapy, etc.)
___ immunosuppression (recent surgery, illness, etc.)   ___ neurological issues

Date symptoms started: _______________ morning ____ afternoon ____

When is discomfort most noticed now: _____ morning

_____ afternoon _____ both

Day(s) of the week most noticed: ___________________________________

Is there an odor noticed: _______ Describe: _______________________________

Do you associate the complaint with any particular event taking place in the vicinity of your work area, such as weather, deliveries, manufacturing process, etc.? If so, describe event: ___________________________________

____

____

What area(s) of the building do you experience the discomfort most:

_______ Have you seen a doctor? Diagnosis?: _______________________________

FOR OFFICE USE ONLY

FILE NUMBER: _______ RECEIVED BY: _______ DATE: _______

Source: NIOSH/EPA BUILDING AIR QUALITY HANDBOOK, Chapter 6, reference 4. Adapted by permission.

INDOOR AIR QUALITY DIARY FORM

DATE: _______________________________

Please fill in this form and return it to IAQ Program Designated Person

NAME: ____________________________ PHONE: ________________________________

BLDG: ___________ DEPT: ____________________ AREA/ROOM: ___________

Please help us to help you by keeping this form near your work area and noting when you experience the conditions that trigger your particular IAQ related symptoms, or when the odor is noticed, if there is one.

DATE	TIME	SEVERITY	COMMENTS

FOR OFFICE USE ONLY

FILE NUMBER: _______ RECEIVED BY: _______ DATE: _______

Source: NIOSH/EPA BUILDING AIR QUALITY HANDBOOK, Chapter 6, reference 4. Adapted by permission.

IAQ COMPLAINT LOG

DATE	PERSON	BLDG	REPAIR DATE	COMMENTS

Source: NIOSH / EPA BUILDING AIR QUALITY HANDBOOK, Chapter 6, reference 4. Adapted by permission.

Appendix E. IAQ Program Design Guide

Wayne Hansen, PE, REA, CEM

SECTION ONE: BASIC INFORMATION

A. IAQ MANAGEMENT TEAM

The following departments and individuals are members of the team responsible for the management and implementation of this program:

1. Representing Administration: Risk Manager

---

2. Representing Safety: Employee Safety Director

---

3. Representing Human Resources: Director

---

4. Infection Control Committee: Manager

---

5. Representing Facilities: Chief Engineer (Designated Person)

B. GENERAL DESCRIPTION OF BUILDINGS

Reduced scale floor plans of each building are shown in the Appendix, Tab A. Appendix Tab B shows the location of MSDS sheets for any product used in the buildings requiring the posting of MSDS sheets.

HOURS	PRIMARY USE	TYPE OF CONSTRUCTION

C. RELATED DOCUMENTS

This health care organization has developed program manuals for in-house use. The following manuals are of interest to this document:

1. *HVAC System Program.* This document details each piece of HVAC and industrial ventilation equipment on each building, and maintenance cycles and log sheets. Where outside contractors are involved, the nature of the involvement, and frequency of that involvement are shown.

2. *Employee Safety Policy and Procedures.* This manual specifies standard safety practices.

3. *Comprehensive IAQ Program Manual.* This manual will contain the report of the initial investigation, recommendations, and air testing results. The follow-up quarterly testing results and simplified report will be included under the appropriate tabs.

SECTION TWO: IAQ PROGRAM IMPLEMENTATION

A. *Records of employee complaints:* Employee complaints related to IAQ issues will be maintained in the Comprehensive IAQ Program Manual. Form D–1 will be filled out for each employee making a complaint, and placed in the appropriate section of that document. Where more in-depth interview information or follow-up information is needed, Form D–2 will be used, along with an Individual IAQ Diary, Form D–3. Form D–4 is a single line entry summary log sheet of each complaint showing the name of the complainant, the date of complaint, building, department, and date complaint resolved. This sheet will serve as a quick reference tracking sheet, and is shown in Appendix D.

B. *IAQ inspections and maintenance related to the HVAC system:* Service and equipment or system inspections records related to IAQ issues will be maintained in the Comprehensive IAQ Program Manual. Form C–1 will be filled out for each HVAC system in each building, and placed in the appropriate section of that document. Form C–2 is a single line entry summary log sheet of each inspection or service showing the building, the HVAC system, the date of performance, area served, and notations. This sheet will serve as a quick reference tracking sheet and is shown in Appendix C.

C. *Comprehensive IAQ program:* This health care organization will establish a program consisting of the following action steps:

1. A physical examination of each building, the current use, and a comparison to the original plan use.

2. An engineering survey of all of the air handling equipment in each building, including an examination of the internal surfaces of the air handling units and ductwork, and contact sampling for surface mold and bacteria. This contact sampling will be performed at the cooling coil, and one section of downstream supply air ductwork. Photographs will be taken as required to document the condition of the systems.

3. Airborne testing of the air within the normally occupied space will be performed in each building to determine the type and level of contamination. The following tests will be performed in each building:

a. Respirable dust.

b. Volatile organic compounds (VOC's) as determined by each area of chemical use.

c. Carbon dioxide (CO_2).

d. Airborne bacteria.

e. Airborne environmental mold.

f. Temperature and coincidental relative humidity.

g. Ozone (O_3).

h. Hydrogen sulfide (H_2S).

i. Radon.

All of the above tests will be performed in locations on each occupied floor of each building relative to each air handling system, with the exception of radon, which will be tested in one location on the ground level of each building.

Some of the above tests may be coincidental with Employee Right to Know Safety Testing. In those cases, copies of those results will be included in this section. Each of the below listed tests will only be performed in central plant areas, manufacturing areas, or laboratories where combustible appliances are located.

j. Carbon monoxide (CO).

k. Oxygen (O_2).

l. Combustible gases (LEL).

One additional set of all tests with the exception of radon will be taken outside to determine the ambient levels.

4. Specific volatile organic compound testing. Chemicals necessary to the function of each area, as shown in the MSDS sheets in the appendix of this document, will be tested for in the initial survey in each area or building where they are

being used. This testing will be coordinated with the Safety Department for use with the employee right to know requirements.

5. A report listing the facility description, tests conducted, recommendations, and all supporting documents including test data, photographs and other materials will be provided. This report will form the initial basis of the Comprehensive IAQ Program Manual.

6. Follow up monitoring. Double blind testing of items 3.a through 3.l With the exception of 3.i will be performed quarterly in one area of each floor of each building. Radon (item 3.i) testing will be done annually in one area of the lowest occupied level of each building. These monitoring reports will be included under the specific date tab in the Comprehensive IAQ Program Manual.

D. *Remodeling and renovation:* All work performed in normally occupied spaces will conform to the following conditions:

1. Employees will be notified well in advance of any work being performed in the area, and will be updated as to completion schedule and any changes to the schedule.

2. Construction areas will be completely sealed from other occupied areas of the building.

3. As far as possible, renovation work will be done when the building is unoccupied.

4. Power tools for cutting, sanding, and drilling will have vacuum hose connections to portable HEPA vacuums.

E. *Education and training:* All employees will be informed of the IAQ program, and their responsibilities to promote good IAQ.

Personnel will avail themselves of educational programs dealing with IAQ in the workplace. Training materials will be kept on file in the Facilities Engineering Department as an educational resource for all employees.

F. *Record keeping:* All records dealing with IAQ will be maintained in the manuals shown in the Related Documents section of this Policy. The Designated Person will make them available to any employee to review upon request.

152

Index

A

D

E

F

G

H

I

P

R

S

We'd like to know what you think of *A Guide to Managing Indoor Air Quality in Health Care Organizations*. Your comments will help us evaluate and improve the value of this publication. Please take a few minutes to give us your opinion and mail this postage-paid card.

Please indicate whether you agree or disagree with the following statements:

	Strongly Agree	Somewhat Agree	Agree	Somewhat Disagree	Strongly Disagree
• *Managing Indoor Air Quality* provides useful and timely information.	❑	❑	❑	❑	❑
• *Managing IAQ* presents new concepts/practical strategies for staying in continuous compliance with EC standards.	❑	❑	❑	❑	❑
• *Managing IAQ* is a value for its price.	❑	❑	❑	❑	❑
• I would recommend *Managing IAQ* to a colleague.	❑	❑	❑	❑	❑

Please rate the value of the following elements:

	Extremely Valuable	Very Valuable	Somewhat Valuable	Not Very Valuable	Not at all Valuable
Overall Layout:	❑	❑	❑	❑	❑
• Chapter 1: Sick Building Syndrome...	❑	❑	❑	❑	❑
• Chapter 2: Building-Related Illness...	❑	❑	❑	❑	❑
• Chapter 3: Risk Factors for Occupational Exposures...	❑	❑	❑	❑	❑
• Chapter 4: Patient Impact	❑	❑	❑	❑	❑
• Chapter 5: Hospital and Employee Impact	❑	❑	❑	❑	❑
• Chapter 6: Developing the IAQ Program	❑	❑	❑	❑	❑
• Chapter 7: Implementing an IAQ Program...	❑	❑	❑	❑	❑
• Chapter 8: Economic Considerations...	❑	❑	❑	❑	❑
• Chapter 9: IAQ Program Case Studies	❑	❑	❑	❑	❑
— Torrance Memorial Medical Center	❑	❑	❑	❑	❑
— Scottsdale Memorial Hospitals	❑	❑	❑	❑	❑
— Kaiser Foundation Hospital	❑	❑	❑	❑	❑
• Appendix A. Indoor Air Quality Management...	❑	❑	❑	❑	❑
• Appendix B. Fungal Growth Checklist	❑	❑	❑	❑	❑
• Appendix C. Facility Mechanical System...	❑	❑	❑	❑	❑
• Appendix D. Log Sheets	❑	❑	❑	❑	❑
• Appendix E. IAQ Program Design Guide	❑	❑	❑	❑	❑
• Index	❑	❑	❑	❑	❑
• Other (please specify) ______________________	❑	❑	❑	❑	❑

I will use *Managing Indoor Air Quality* to (please check all that apply):

❑ Learn more about a subject I know little about   ❑ Educate staff

❑ Help facilitate ongoing performance improvement   ❑ Other (please specify)_______________________________

Characteristics of my organization:
- My title is: ___________________________________
- Type of organization (such as, surgery center, student health services): _______________________________
- Size: ❑ Small ❑ Medium ❑ Large

Please make any additional comments about *Managing Indoor Air Quality*: _______________________________

____

If we may contact you about this or other Joint Commission products, please provide the following information:

Name: ___________________________________ Telephone:_______________________________

Organization: ____

Street Address:____

City, State, ZIP: ____

If you would like information about Joint Commission publications, please call the Customer Service Center at 630/792-5800. **Thank you!**

BUSINESS REPLY MAIL
FIRST CLASS MAIL   PERMIT NO 632   VILLA PARK IL

POSTAGE WILL BE PAID BY ADDRESSEE

JOINT COMMISSION ON ACCREDITATION
OF HEALTHCARE ORGANIZATIONS
ATTN KRISTINE TOMASIK
ONE RENAISSANCE BOULEVARD
OAKBROOK TERRACE IL 60181-9887

NO POSTAGE
NECESSARY IF
MAILED IN THE
UNITED STATES